Survival Guide for
Introductory Chemistry
With Math Review

Charles H. Atwood
University of Georgia, Athens

BROOKS/COLE
CENGAGE Learning™

Australia • Brazil • Japan • Korea • Mexico • Singapore • Spain • United Kingdom • United States

For product information and technology assistance, contact us at **Cengage Learning Customer & Sales Support, 1-800-354-9706**

For permission to use material from this text or product, submit all requests online at **www.cengage.com/permissions** Further permissions questions can be emailed to **permissionrequest@cengage.com**

ISBN-13: 978-0-495-82826-6
ISBN-10: 0-495-82826-2

Brooks/Cole
10 Davis Drive
Belmont, CA 94002-3098
USA

Cengage Learning is a leading provider of customized learning solutions with office locations around the globe, including Singapore, the United Kingdom, Australia, Mexico, Brazil, and Japan. Locate your local office at: **www.cengage.com/international**

Cengage Learning products are represented in Canada by Nelson Education, Ltd.

To learn more about Brooks/Cole, visit **www.cengage.com/brookscole**

Purchase any of our products at your local college store or at our preferred online store **www.ichapters.com**

Printed in the United States of America
1 2 3 4 5 6 7 12 11 10 09 08

Table of Contents

iii

Module 5
Chemical Nomenclature

Module 6
Understanding Chemical Formulas and Using the Mole Concept

Module 7
Understanding Chemical Reaction Equations

Module 8
Chemical Reactions In Aqueous Solutions

Module 9
Chemical Reaction Stoichiometry

Module 10
Electronic Structure of Atoms

Module 11
Periodic Trends in Chemistry

Module 12
Chemical Bonding

Module 13
Molecular Shapes and Polarity

Module 14
Gases

Module 15
Liquids and Solids

Module 16
Solutions

Module 17
Acids and Bases

Module 18
Chemical Equilibrium

Module 19
Oxidation-Reduction Reactions and Electrochemistry

Module 20
Nuclear Chemistry

Math Review

Preface

To The Student:

This is the third version of a Survival Guide for various chemistry courses that I have written in the last five years. I want to assure you that CENGAGE would not have me write a third one if the first two had not been successful in helping students understand the key topics in a typical chemistry course. If you will use this guide as a method to help you in preparing for your tests in your introductory chemistry course, it will pay dividends. I have tried to write it as if you were sitting beside me at my desk and I was helping you solve the problems. Consequently, you will see solved problems with numerous arrows and boxes pointing where numbers, equations, and other pertinent information come from and are introduced into the problems. Included are many of the most commonly asked questions in a typical introductory chemistry course. The essence of such a course has been distilled into a small set of topics that are included in this survival guide. The **INSIGHT** and **CAUTION** boxes are designed to draw your attention to important details that will help you succeed. This guide will assist you in studying for tests at your institution. I sincerely believe that using this survival guide will aid your comprehension and success in introductory chemistry at nearly every institution of learning.

Acknowledgements:

Once again, my CENGAGE colleagues, Lisa Lockwood, Elizabeth Woods, Lisa Weber, and Teresa Trego, have been instrumental in the production of this **Survival Guide**. Their advice and guidance have insured that I made this survival guide more useful for the students. I am fortunate to have such wonderful colleagues in the publishing business.

This Survival Guide was reviewed by Mark Erickson from Hartwick College. His due diligence caught many typos, mistakes, misstatements, missing commas, and other editorial details that I managed to miss. His assistance in the work is greatly appreciated.

My wife, Judy, gets to see me disappear upstairs to my home office nearly every evening to work on this and other projects. She never ceases to amaze me with her love for me and her willingness to let me work on various projects at the expense of her time with me. I cannot adequately express in words my love and appreciation to her.

I dedicate this Survival Guide to the memory of my mother and father, Margaret and Garland Atwood. The two of them raised three great kids in some very difficult financial and emotional times. Somehow they managed to keep all three of us focused on the important things in life, kept us fed, clothed, and determined to succeed. I want you both to know that your hard work has borne a rich bounty of fruit. I love you both for what you gave us, the best of yourselves.

Module 1
An Approach to Problem Solving

Introduction

Almost from the moment that you begin your introductory chemistry course, you will discover that chemists prize problem solving skills. Problem solving is how most chemists earn their living. Consequently, they believe that chemistry students must come to appreciate how to solve problems. Your instructor may well show you a set of problem solving steps which they prefer. If so, then use those steps. Here is a set that have worked for many students which should help improve your success in this class.

Module 1 Key Concepts
Problem Solving Check List
1. **Carefully read the entire problem.**
2. **Make sure you know <u>precisely</u> what the problem is asking you.**
   a) Frequently the question will be a sentence that ends with a question mark. Read that sentence two or three times, if necessary, until you are certain you know precisely what the problem is asking.
   b) If the problem includes the solution's expected units (for example m, ft, ^{o}C, or gal) utilize that as a hint to solve the problem. Use them in your problem solving strategy.
3. **Find the chemical principle which the problem is based upon.**
   a) Every problem is based upon something that you were taught in class. It may not be apparent initially, but it is there.
   b) After you identify the chemical principle, use the techniques that you were taught in class.
4. **Determine which parts of the problem are relevant to the solution.**
   a) Chemistry problems frequently have more information than is necessary to solve the problem.
      i. This is not to confuse you but because to chemists the problem is incorrect without that information.
      ii. For example, the temperature will often be included whether it is relevant to the solution or not. To chemists, the relevant parts of the problem have no physical reality without knowing the value of the temperature.
   b) Ignore the irrelevant parts and focus on the relevant parts.
5. **Map out a solution pathway.**
   a) If necessary, write down a solution pathway. As your problem solving skills improve, you may do this without writing it down.
   b) Dimensional analysis is one method to map the solution pathway.
6. **Check your solution.**
   a) After you have numerically solved the problem, check the solution.
   b) Does the solution have the correct units?
   c) Is the numerical value reasonable?

Gas Problem Solving Sample Exercise

1. *A gas sample at 37.0°C has a volume of 12.0 L under a pressure of 2.50 atm. What is the volume of the gas in L if the pressure is changed to 0.500 atm?*

   1. Read the problem

   2. Understand precisely what the problem is asking

   a) This problem is asking us to determine the volume of a gas, in L, as its pressure is changed at constant temperature.

   3. Identify the chemical principles

   a) The chemical principle involved, Boyle's Law, explains changes in gas volume or pressure at constant temperature.

   4. Determine the relevant pieces of information

   a) Boyle's Law is only applicable at constant temperature.

   b) We shall not use the 37.0°C in solving the problem, it is an irrelevant piece of information in the mathematical solution, but it tells us that Boyle's Law is applicable.

   5. Create a solution pathway

   a) In mathematical form Boyle's Law is $P_1V_1 = P_2V_2$. Our problem gives values for P_1, V_1, and P_2.

   b) Algebraically solving for V_2 gives $V_2 = \dfrac{P_1V_1}{P_2}$ which is the solution pathway.

   c) Checking the problem's units show that $V_2 = \dfrac{P_1(\text{atm})V_1(\text{L})}{P_2\,(\text{atm})}$ and the atm units cancel leaving the final units of L.

   d) Insert the numbers into the problem.

   $$V_2 = \frac{2.50\ \text{atm}\ 12.0\ \text{L}}{0.500\ \text{atm}} = 60.0\ \text{L}$$

   6. Check your answer

   a) Boyle's law indicates that pressure and volume behave inversely, as a gas's pressure increases the volume decreases and as the gas's pressure decreases the volume increases.

   b) In this problem, the pressure is decreasing from 2.50 atm to 0.500 atm so the volume must increase.

   c) The calculated answer of 60.0 L has the correct units and is reasonable because the volume has increased from 12.0 L to 60.0 L.

Density Problem Solving Sample Exercise

2. *Gold has a density of 19.3 g/mL. How many pounds of gold are required to fill a Coca-Cola® can? The can has a volume of 355 mL.*

   1. Read the problem

   2. Understand precisely what the problem is asking

   a) The problem is asking us to determine the mass of gold, in lb, that is required to fill a Coca-Cola® can.

   3. Identify the chemical principles

   a) Density and volume are the chemical principles involved.

4. Determine the relevant pieces of information

a) There are no irrelevant pieces of information given in this problem.

b) However, one piece of information that is missing is that there are 454 grams in a pound. It is not unusual for your instructor to either expect you to know this fact or for it to be included in an information table.

5. Create a solution pathway

a) In mathematical form $D = \dfrac{m}{V}$. The problem gives values for D and V.

b) Algebraically solving for m gives $m = D \times V$ which is part of the solution pathway.

c) After we have determined the mass in grams, we will use dimensional analysis to determine the mass in pounds as follows.

$$? \text{ lb} = m(\cancel{g})\frac{1 \text{ lb}}{454 \cancel{g}}$$

d) Insert the numbers into the problem.

$$? \, m = D \times V = 19.3 \, \frac{g}{\cancel{mL}} \times 355 \, \cancel{mL} = 6.85 \times 10^3 g$$

$$? \, \text{lb} = 6.85 \times 10^3 \, \cancel{g} \, \frac{1 \text{ lb}}{454 \, \cancel{g}} = 15.1 \text{ lb}$$

6. Check your answer

a) Density is the conversion factor between mass and volume. Once we know the density and volume, we can solve for the mass.

b) The units of this problem are correct.

c) Gold is a very dense metal so it is not surprising that such a small volume has the large mass of 15.1 lb.

Dimensional Analysis Problem Solving Sample Exercise

3. *Pennies have a mass of 2.50 g. If the penny is made entirely of copper, how many copper atoms are in one penny?*

   1. Read the problem

   2. Understand precisely what the problem is asking

   a) The problem is asking you to determine the number of copper atoms in 2.50 g of copper.

   3. Identify the chemical principles

   a) The mole concept is the chemical principle involved in this problem.

   4. Determine the relevant pieces of information

   a) There are no irrelevant pieces of information given in this problem.

   b) To work this problem you will have to look up one important piece of information from the periodic table – the molar mass of copper which is 63.55 g/mol.

   c) You will also have to know an important piece of information - Avogadro's number (6.022×10^{23} atoms/mol).

3

5. Create a solution pathway

a) The solution pathway for this problem requires using the penny's mass and the molar mass of copper to determine the number of moles. Once the number of moles is known, we can use Avogadro's number to determine the number of copper atoms.

b) The number of moles is determined using

$$? \text{ moles of Cu} = m(\text{g of Cu}) \frac{1 \text{ mol of Cu}}{63.55 \text{ g of Cu}}.$$

c) From the number of moles we will determine the number of Cu atoms as follows.

$$? \text{ atoms of Cu} = \text{moles of Cu} \; \frac{6.022 \times 10^{23} \text{ atoms of Cu}}{\text{moles of Cu}}$$

d) Insert the numbers into the problem.

$$? \text{ moles of Cu} = 2.50 \text{ g of Cu} \; \frac{1 \text{ mol of Cu}}{63.55 \text{ g of Cu}} = 0.0393 \text{ mole of Cu}$$

$$? \text{ atoms of Cu} = 0.0393 \text{ mole of Cu} \; \frac{6.022 \times 10^{23} \text{ atoms of Cu}}{\text{mole of Cu}}$$

$$= 2.37 \times 10^{22} \text{ atoms of Cu}$$

6. Check your answer

a) Moles are the conversion factor between mass and number of atoms. If we know the mass, we can solve for the number of atoms.

b) The units of this problem are correct.

c) Atoms are incredibly small particles of matter. In any visible object there are billion upon billions upon billions of atoms. Thus the answer of 2.37×10^{22} atoms is reasonable.

Practice Problems

Start developing your problem solving skills with these three problems.

To help you the correct chemical principle involved is indicated for each problem.

4. *A gas under a pressure of 1.25 atm has a volume of 3.15 L at 25.0°C. What is its volume at 50.0°C?*

    The chemical principle is Charles's Law. Remember to convert temperature to K.

    The correct answer is 3.41 L.

5. *Helium has a density of 1.7×10^{-4} g/mL. What mass of He is contained in a balloon having a volume of 22.4 L at 25.0°C and 1.00 atm of pressure?*

    The chemical principle is density.

    The correct answer is 3.8 g.

6. *The quarter, a US coin equal to \$0.25, contains 5.82×10^{22} Ni atoms. Assuming that quarters are pure Ni, determine the mass of a quarter.*

    The chemical principle is the mole.

    The correct answer is 5.67 g.

Module 2
Metric System, Significant Figures, Dimensional Analysis, and Density

Introduction

This module introduces a) exponential (scientific) notation, b) the basic rules of the metric system and significant figures, c) the use of dimensional analysis to help solve problems including metric to USCS conversions, d) converting temperature from Celsius to Fahrenheit to Kelvin, and e) density problems. These are typical topics introduced in an early chapter of introductory chemistry textbooks.

Module 2 Key Equations & Concepts

1.
$$^{\circ}F = \frac{9}{5}\,^{\circ}C + 32$$

$$^{\circ}C = \frac{5}{9}\left(^{\circ}F - 32\right)$$

These two relationships are used to convert Celsius temperatures to Fahrenheit and vice versa.

2.
$$^{\circ}C = K - 273.15$$

$$K = \,^{\circ}C + 273.15$$

These two relationships are used to convert Celsius temperatures to Kelvin and vice versa.

3.
$$D = \frac{m}{V}$$

The density equation is used to determine:
a. density, when mass and volume are given
b. mass, when density and volume are given
c. volume, when density and mass are given

4. **Metric system**

A system of units using a series of multipliers to convert from one size to another. The table just below gives a common set of prefixes.

Prefix Name	Prefix Symbol	Multiplication Factor
giga-	G	1,000,000,000 or 10^9
mega-	M	1,000,000 or 10^6
kilo-	k	1,000 or 10^3
deci-	d	0.1 or 10^{-1}
centi-	c	0.01 or 10^{-2}
milli-	m	0.001 or 10^{-3}
micro-	μ	0.000001 or 10^{-6}
nano-	n	0.000000001 or 10^{-9}
pico-	p	0.000000000001 or 10^{-12}

Exponential (Scientific) Notation Sample Exercises
1. Write 1,527,349 in exponential notation.
The correct answer is 1.527349×10^6

> There are 6 numbers between the decimal place and the 1 in this number corresponding to 6 powers of 10.

$$6 \text{ powers of 10} \qquad 6 \text{ powers of 10}$$
$$1,\overbrace{527,349} = 1.527349 \times \overbrace{10^6}$$

> 6 powers of 10 written as 10^6.

Because 1,527,349 is a number greater than 1, the powers of 10 are positive.

2. Write 0.00008234 in exponential notation.
The correct answer is 8.234×10^{-5}.

> There are 5 numbers between the decimal place and the 2 in this number corresponding to 5 powers of 10.

$$5 \text{ powers of 10} \qquad 5 \text{ powers of 10}$$
$$0.\overbrace{00008}\;234 = 8.234 \times \overbrace{10^{-5}}$$

> 5 powers of 10 written as 10^{-5}.

Since 0.00008234 is a number less than 1, the powers of 10 are negative.

3. Write 6.23×10^{-7} in decimal form.
The correct answer is 0.000000623.

$$7 \text{ powers of 10} \qquad 7 \text{ powers of 10}$$
$$6.23 \times \overbrace{10^{-7}} = 0.\overbrace{0000006}23$$

Because the powers of 10 are negative, the number is less than 1. Ensure that there are 7 powers of 10 (the 6 and six zeroes) between the decimal place and the 2.

Exponential (scientific) notation is a compact method
of writing very large or small numbers.

Metric System Sample Exercises
4. How many mm are there in 3.45 km?
The correct answer is 3.45×10^6 mm.

The table indicates that there are 1000 m in 1 km and that 1 mm = 0.001 m.

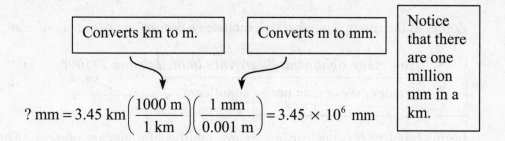

One way to help remember how to work these problems correctly is to notice which of the units is the largest. In this problem, the km is a much larger unit than the mm. Thus we reasonably expect that there will be many of the smaller unit, mm's, in the large units. Notice that the answer indicates there are 3.45 million mm in 3.45 km, which is sensible.

5. How many mg are in 15.0 pg?

The correct answer is 1.5×10^{-8} mg.
From the table we see that 1 pg = 10^{-12} g and 1 mg = 10^{-3} g.

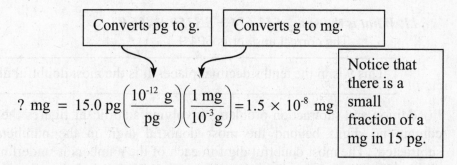

In this problem picograms, pg, are the smaller unit. We should expect that there are very few milligrams, mg, in 15.0 pg. Consequently, the answer 1.5×10^{-8} mg is reasonable.

Significant Figures Sample Exercises

6. How many significant figures are in the number 58062?

The correct answer is five significant figures.

This zero is significant because it is embedded in other significant digits.

7. How many significant figures are in the number 0.0000543?

The correct answer is three significant figures.

These zeroes are not significant because their purpose is to indicate the position of the decimal place.

8. How many significant figures are in the number 0.009120?

The correct answer is four significant figures.

These three zeroes are not significant just as in exercise 7.

This zero is significant.

7

Zeroes at the end of a number that includes a decimal point are significant.

9. How many significant figures are in the number 24500?

> These zeroes may or may not be significant.

In this problem it is unclear how many significant digits are present. There could be as few as three (the 2, 4, and 5) or as many as five if both of the zeroes are significant.

10. How many significant figures are in the number 2.4500×10^4 ?
The correct answer is five significant figures.

> As written both of these zeroes are significant.

Notice that this number is the same as in exercise 9 but written in scientific notation. Furthermore, it uses the rule displayed in exercise 8 as well. Thus the significant digits are the 2, 4, 5, and both of the zeroes.

11. What is the sum of 12.674 + 5.3150 + 486.9?
The correct answer is 504.9.

> This 9 is in the tenths decimal place. It is the most doubtful digit in the sum.

In addition and subtraction problems involving significant figures the final answer must contain no digits beyond the most doubtful digit in the numbers being added or subtracted. The most doubtful digit in each of the numbers is underlined 12.67$\underline{4}$, 5.315$\underline{0}$, 486.$\underline{9}$. Notice that the 486.9 has the most doubtful digit because the 9 is only in the tenths position and the other numbers are doubtful in the thousandths (12.674) and ten thousandths (5.3150) positions. The final answer must have the final digit in the tenths position.

12. What is the correct answer to this problem, $2.6138 \times 10^6 - 7.95 \times 10^{-3}$?
The correct answer is 2.6138×10^6.

> This 8 is the most doubtful digit in the sum. It is in the hundreds position.

The number 2.6138×10^6 can be also written as 2,613,800. Its most doubtful digit, the 8, is in the hundreds position. The other number, 7.98×10^{-3}, can be written as 0.00795. Its most doubtful digit, the 5, is in the one millionths position. Consequently, the final answer cannot extend beyond the 8 in 2.6138×10^6. ***When adding and subtracting two numbers, both numbers must be expressed to the same power of 10 to determine the most doubtful digit.***

13. What is the correct answer to this problem, 47.893 × 2.64 ?
        The correct answer is 126.

> This number contains only 3 significant digits. It determines the final result.

In multiplication and division problems involving significant figures, the final answer must contain the same number of significant figures as the number with the least number of significant figures. In this problem, 47.893 has five significant figures and 2.64 has three significant figures. The correct answer must have three significant figures to match the number of significant figures in 2.64, thus the answer is 126.

14. What is the correct answer to this problem, $1.95 \times 10^5 \div 7.643 \times 10^{-4}$?
        The correct answer is 2.55×10^8

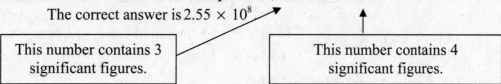

> This number contains 3 significant figures.

> This number contains 4 significant figures.

Just as in exercise 10, the number with fewest significant digits, 1.95×10^5 having three significant digits, determines the final answer of 2.55×10^8 which also has three significant digits.

Metric to USCS Sample Exercises
        In chemistry we often perform calculations that require changing from one set of units, say ft or in or lb, to a second set of units like Mm or km or kg. A convenient method to convert units without making arithmetic errors is called dimensional analysis. In dimensional analysis common conversion factors, frequently given in your textbook, are arranged so that one set of units cancels, converting the problem to the second, third, fourth, and so forth sets of units. In this next problem, let's practice the problem solving check list from Module 1.

15. How many Mm are in 653 ft?
        The correct answer is 1.99×10^{-4} Mm.

1. **Read the problem**
   a) Reading this problem carefully is important because many students will convert feet to millimeters, mm, not Mm.
2. **Understand precisely what the problem is asking**
   a) The problem is asking us to convert 653 feet to megameters, a metric unit equivalent to 10^6 meters.
3. **Identify the chemical principles**
   a) Dimensional analysis, the metric system, and the USCS system are the chemical principles in this problem.
4. **Determine the relevant pieces of information**
   a) 653 feet and Mm are relevant pieces of information.
   b) You must also know or have an information sheet that tells you the number of inches in a cm, or the number of feet in a m, etc.

9

c) It is a very good idea to know at least one conversion factor from the metric to USCS system and vice versa.

d) Generally, your instructor will expect you to know conversions from feet to inches, etc. and conversions within the metric system such as given in the table at the start of this Module.

5. Create a solution pathway

a) We must get from a USCS unit (feet) to a metric unit (Mm).

b) If you know that 1 in = 2.54 cm, the following solution pathway is appropriate ft → in → cm → m → Mm.

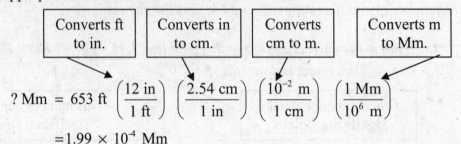

$$? \text{Mm} = 653 \text{ ft} \left(\frac{12 \text{ in}}{1 \text{ ft}} \right) \left(\frac{2.54 \text{ cm}}{1 \text{ in}} \right) \left(\frac{10^{-2} \text{ m}}{1 \text{ cm}} \right) \left(\frac{1 \text{ Mm}}{10^6 \text{ m}} \right)$$

$$= 1.99 \times 10^{-4} \text{ Mm}$$

c) Notice that the problem is arranged so that each successive conversion factor makes progress in the solution pathway.

6. Check your answer

a) Each conversion factor is correct.

b) Conversion factors are arranged so that the units cancel.

c) Feet are small to medium-sized units while Mm are large units. It makes sense that 653 feet are a small fraction of a Mm.

16. Low flow toilets commonly use 1.5 gallons of water per flush. How many mL of water are consumed in one flush?

The correct answer is 5.7×10^3 mL.

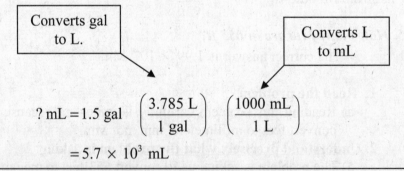

$$? \text{mL} = 1.5 \text{ gal} \left(\frac{3.785 \text{ L}}{1 \text{ gal}} \right) \left(\frac{1000 \text{ mL}}{1 \text{ L}} \right)$$

$$= 5.7 \times 10^3 \text{ mL}$$

Both 3.785 and 1000 are conversion factors with several significant figures but because the problem uses 1.5 gallon the final answer must be 2 significant figures.

10

17. How many grams of fertilizer are in a 40.0 lb bag of fertilizer?
The correct answer is 1.81×10^4 g.

Converts lb to kg		Converts kg to g

$$? \; g = 40.0 \; lb \; \left(\frac{1 \; kg}{2.205 \; lb} \right) \left(\frac{1000 \; g}{kg} \right)$$

$$= 1.81 \times 10^4 \; g$$

Temperature Conversion Sample Exercises

18. In the dark areas on the surface of the Earth's Moon the temperature can be as low as -184°C. What is this temperature in Kelvin (K) and Fahrenheit (F)?
The correct answer is 89 K and -299°F.

From the Key Equations and Concepts use the centigrade to K relationship. In this subtraction exercise, significant figures are determined by the 4 in -184.

$$? \; K = -184°C + 273.15 = 89.15 \; K \approx 89 \; K$$

From the Key Equations and Concepts Use the relationship to convert centigrade to Fahrenheit. In this multiplication and addition exercise, the number of significant figures is also decided by the 4 in -184.

$$? \; °F = \frac{9}{5}(-184°C) + 32 = -299°F$$

Density Sample Exercises

19. What is the mass, in g, of a 68.2 cm³ sample of ethyl alcohol? The density of ethyl alcohol is 0.789 g/cm³.
The correct answer is 53.8 g.

Density is the conversion factor between volume and mass of a substance.

$$? \; g = 68.2 \; cm^3 \left(\frac{0.789 \; g}{1 \; cm^3} \right)$$

$$= 53.8 \; g$$

Final units are g because the cm³ in the density cancels with the original volume.

20. What is the volume, in cm³, of a 237.0 g sample of copper? The density of copper is 8.92 g/cm³.

The correct answer is 26.6 cm³.

Density, inverted, cancels the original mass, in g, leaving the volume in cm³.

$$? \, cm^3 = 237.0 \, g \left(\frac{1 \, cm^3}{8.92 \, g} \right)$$

$$= 26.6 \, cm^3$$

21. What is the density of a substance having a mass of 25.6 g and a volume of 74.3 cm³?

The correct answer is 0.345 g/cm³.

$$? \, g/cm^3 = \frac{25.6 \, g}{74.3 \, cm^3} = 0.345 \, g/cm^3$$

Density's units, g/cm³, help determine the correct order of division.

Module 3
Matter and Energy

Introduction

This module introduces some of the most basic concepts in chemistry including <u>a) states of matter; b) physical and chemical properties and changes; and c) the difference between elements, compounds, substances, and mixtures</u>. These topics form the basis of many of the subsequent chapters in most introductory chemistry textbooks.

Module 3 Key Equations & Concepts

1. **Solid, Liquid, and Gas – The Three Most Common States of Matter**

   All substances (sugar, water, and helium for example) exist in one of the three common states of matter at room temperature and pressure. Each state can be transformed to another state by heating or cooling. For example, liquid water can be cooled to form solid ice or heated to form gaseous steam.

2. **Physical Properties and Changes**

   When a substance is transformed without changing its chemical identity a physical change has occurred. For example, heating ice will turn it first into liquid then gaseous water. This is a physical change because water retains its chemical identity (remains water) throughout the process. Physical properties are measurements of matter that do not change the substance's chemical identity. Some examples of physical changes are:
   - **a)** evaporation
   - **b)** density
   - **c)** color

3. **Chemical Properties and Changes**

   Transformations that change a substance's chemical identity are chemical changes. For example, octane, a component of gasoline, burns in the presence of oxygen to form water and carbon dioxide. In this example, octane is changed from a compound containing carbon and hydrogen atoms to two new compounds containing hydrogen, oxygen, and carbon atoms. Chemical properties are measurements of matter that indicate how the substance's chemical identity has changed. Some examples of chemical changes are:
   - **a)** decomposition
   - **b)** combustion
   - **c)** polymerization

4. **Elements, Compounds, Substances, and Mixtures**
   - **a)** Elements are composed of atoms all having the same chemical identity. There are 111 known elements. Examples include C, N, H, and O.
   - **b)** Compounds are composed of atoms having two or more atoms of different elements. There are millions of known chemical compounds. Examples include water, sugar, table salt, and baking soda.

---

c) Substances are chemical species that are present in pure form. In a pure substance there are only atoms or molecules of one element or atoms, molecules or formula units of one compound. There are millions of known substances. Examples include aluminum foil, baking soda, and sugar.

d) Mixtures are collections of elements or substances. Mixtures can be homogeneous (no distinction between the components of the mixture) such as tap water, cooking oil, air, or salt water. There are also heterogeneous mixtures (in which the components of the mixture are distinguishable) such as a hot fudge sundae, thousand island salad dressing, or pickle relish.

Notice that we can use the problem solving steps check list to help solve this problem. It could also be used for the other problems in this module.

States of Matter Sample Exercises

1. Shown below are atomic/molecular representations of solid and gaseous carbon dioxide as well as solid and liquid sodium. Associate each drawing with the correct state and substance.

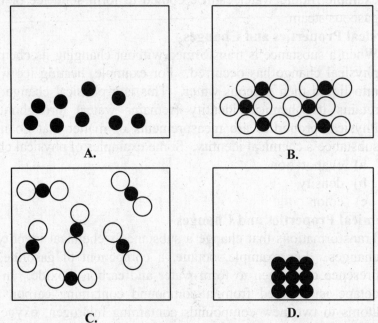

The correct answer is image A. represents liquid sodium, image B. represents solid carbon dioxide, image C. represents gaseous carbon dioxide, and image D. represents solid sodium.

1. **Read the problem**
   a) For this problem you must read not only the words but pay careful attention to the images as well.
2. **Understand precisely what the problem is asking**
   a) The problem asks us to associate images with physical states of matter and atomic/molecular structure of Na and CO_2.
3. **Identify the chemical principles**
   a) The chemical principles involved are to understand the structure of atoms, ions, and molecules and recognize three physical states of matter.

14

4. Determine the relevant pieces of information
a) In this problem the images combined with the question phrasing are the important pieces of information. There are no irrelevant pieces of information.

5. Create a solution pathway
a) Distinguish between solids, liquids, and gases.
1) Solids are the most ordered and closely packed state of matter.
2) Liquids are more random and less closely packed than solids.
3) Gases are the least ordered and most dispersed of these thee states of matter.

b) Distinguish between the atomic/molecular structure of Na and CO_2.
1) CO_2 is a molecule made of three atoms (1 C and 2 O atoms). Its symbolic representation must be ⃝●⃝.
2) Na is a substance composed of individual atoms. Its symbolic representation must be ●.

c) Assemble the information in a logical pattern.
1) Images B. and D. indicate solids because the particles constituting the substances are highly ordered and close to one another.
2) Image B. is carbon dioxide because each unit in the image has three atoms, ⃝●⃝, corresponding to the three atoms in CO_2.
3) Image D. is the solid sodium because there are single particles stacked in an ordered arrangement.
4) Image C. is the gaseous carbon dioxide. Note the molecules are randomly distributed inside the box.
5) Image A. is the liquid sodium. Single particles are distributed inside the box but not as far apart as is the case for gaseous carbon dioxide.

6. Check your answer
a) We have identified the solids as being more ordered with the particles close together.
b) We have identified molecular CO_2 and atomic Na.
c) We have recognized that gaseous CO_2 is farther apart and more random than liquid Na.
d) All of the answer options have been identified and are sensible based upon our understanding of atomic/molecular theory.

> One of the key clues to understanding chemistry is the ability to visualize what chemicals look like if we could see atoms, ions, or molecules. Work on developing this ability as you progress through the course.

Physical and Chemical Properties Sample Exercise
2. Classify each of the following properties as either physical or chemical.
a. A mailbox rusting in the sun.
b. Salt dissolving in water.

> c. *Dry ice disappearing into air.*
> d. *Sulfur is a yellow colored solid at room temperature.*
> e. *Sulfur burns in air with a blue-colored flame.*
>   The correct answer is a and e are chemical properties while b, c, and d are physical properties.

In b, c, and d the chemical substances (salt, dry ice, and sulfur) retain their chemical identities. Salt is salt before and after it dissolves in water. Dry ice (solid carbon dioxide) becomes gaseous carbon dioxide as it disappears into air. Sulfur remains sulfur as a solid. The steel that a mailbox is made of reacts with air in the presence of moisture to form iron oxide, a vastly different chemical compound from steel. Sulfur burns in air to form sulfur dioxide which is chemically different from sulfur.

Physical and Chemical Changes Sample Exercise
> 3. *Classify each of the following changes as either physical or chemical.*
>   a. *Orange colored potassium dichromate crystals dissolve in water to form an orange colored solution.*
>   b. *Plants absorb carbon dioxide and release oxygen in their life cycle.*
>   c. *Natural gas burns in air releasing carbon dioxide, water, and heat.*
>   d. *A steak cooking on the grill.*
>   e. *Purple colored iodine crystals from a purple colored gas when heated.*
>   The correct answer is a and e are physical changes while b, c, and d are chemical changes.

In b, c, and d, the chemical substances (carbon dioxide, natural gas, and especially the steak's exterior) change their chemical identities. Carbon dioxide changes into sugars, oxygen, and water inside the plant. Natural gas (primarily methane, CH_4) becomes gaseous carbon dioxide and water vapor as it burns in air. As it is cooked, the steak's exterior changes from a tough, chewy muscular substance into a tender, moist and flavorful dish. Potassium dichromate is still potassium dichromate whether it is dissolved in water or not. Solid or gaseous iodine is still iodine and has not changed its chemical identity.

> 4. *Classify each of these images as either physical or chemical changes.*

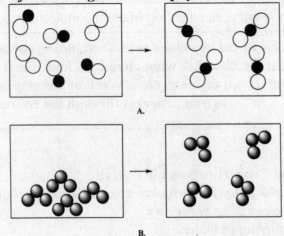

A.

B.

16

The correct answer is image A is a chemical change and image B is a physical change.

Image A indicates that the ⬭ molecules react with ⬭ molecules to form ⬭ molecules. Notice that atoms move from the ⬭ to the ⬭ forming a new substance, ⬭. The exchanging of atoms is indicative of a chemical reaction. Image B is a physical change because the ⬭ molecules retain their chemical identity. There is no exchange of atoms from the first box to the second.

Pure Substances and Mixtures Sample Exercise
5. *Classify each of the following as either pure substances or mixtures.*
    a. *Beach sand*
    b. *Aluminum foil*
    c. *A wedding ring*
    d. *An oak leaf*
    e. *A diamond*

    The correct answer is b and e are pure substances while a, c, and d are mixtures.

Aluminum foil is composed solely of aluminum metal (if we ignore the very thin oxide coating on its surface). A diamond is composed solely of carbon atoms and is also a pure substance. Beach sand is a complex mixture of several silicon oxides, various inorganic salts, and tiny bits of organic matter from the fish, crabs, and birds that live in or over the sand. In the United States, wedding rings are alloys, solid mixtures, made of gold, silver, platinum, zinc, and a few other trace metals. Oak leaves are also mixtures of several complex carbohydrates and other substances that make the cells and surface of the leaves.

Elements, Compounds, Pure Substances and Mixtures Sample Exercise
6. *Classify each of the following images as representing elements, compounds, pure substances, or mixtures.*

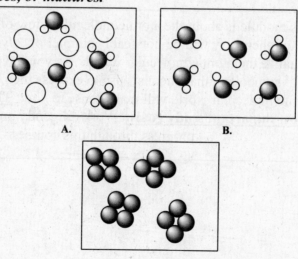

A.   B.

C.

17

The correct answer is image A represents a mixture, image B represents a compound and a pure substance, image C represents an element and is also a pure substance.

Notice that in B and C each representation of a molecule is the same as the other ones in the box. This is indicative of pure substances. Image A on the other hand has some subunits composed of three atoms, ⚬⚬⚬, and others composed of single atoms, ○, indicative of a mixture. Image B is a compound because each molecule is composed of 2 ○ atoms and 1 ⚫ atom. (Remember compounds have atoms of more than one element.) Image C. is an element because each molecule in the box is made of the same elements. There are only ⚫ atoms!

Homogeneous and Heterogeneous Mixtures Sample Exercise

7. *Classify each of the following as a pure substance, homogeneous mixture, or heterogeneous mixture.*
   a. *Air*
   b. *Crunchy peanut butter*
   c. *A brass candlestick*
   d. *Baking soda*
   e. *Liquid dishwashing detergent*

   The correct answer is a, c, and e are homogeneous mixtures, b is a heterogeneous mixture, and d is a pure substance.

Baking soda is pure sodium hydrogen carbonate, a pure ionic chemical compound. Crunchy peanut butter is a heterogeneous mixture. The butter is a complex mixture made from crushed peanuts and peanut oil while the whole peanuts make the mixture heterogeneous. Air (composed of nitrogen, oxygen, carbon dioxide, and other gases), brass (an alloy composed of copper, zinc, iron, and several other metals), and liquid dishwashing detergent (made of several cleaning agents and surfactants) are all homogeneous mixtures.

> YIELD
>
> This module is about the atomic/molecular view of the world. Your study of chemistry will be noticeably simpler once you understand how to visualize the atomic/molecular world. As you develop this ability, the symbolism that chemists use to express their view of the atomic/molecular world will make sense to you. **This is one of your most important study goals.** Work diligently on this skill as you progress through this course.

18

Module 4
Elements, Atoms, and the Periodic Table

Introduction

In this module, we look at the elements which make all substances. We also examine atoms and the subatomic particles which comprise them. Finally, we look at how the elements are arranged on the periodic table. The important thing to understand from this module is a) how to find an element on the periodic table, b) determine the number of protons, neutrons, and electrons present in an atom or ion of an element, c) how to calculate an elements' atomic weight, then d) realize that the periodic table has a systematic arrangement based upon proton numbers of the elements. You may need a periodic table as you work through this module.

Module 4 Key Equations & Concepts

1. Proton number = Z

Presently there are 111 known elements. Atoms of every element are composed of protons and neutrons in the nucleus and electrons surrounding the nucleus. What distinguishes one element from the others is the number of protons in the nucleus. The atomic number, found on the periodic table, equals the number of protons.

2. Protons, neutrons, and electrons

Protons are positively charged subatomic particles found in the nucleus. Neutrons are neutral subatomic particles also found in the nucleus. Electrons are negatively charged subatomic particles that surround the nucleus. Every element has a fixed number of protons and electrons (if neutral) but variable numbers of neutrons. Isotopes are atoms of an element having different numbers of neutrons. An element's atomic mass (symbol A) is equal to the proton number (symbol Z) plus the neutron number (symbol N).

$$A = Z + N$$

3. Ions

An atom that has equal numbers of protons and neutrons is electrically neutral, it has no net charge. Because electrons are negatively charged, an atom that has more electrons than protons is a negatively charged ion. An atom that has fewer electrons than protons is a positively charged ion.

4. Atomic weight

Most elements have more than one stable isotope. In a sample of an element these stable isotopes are present in different percentages. The atomic weight is a weighted average of the stable isotopes of an element. It is the number at the bottom of an element's box on the periodic table.

5. Periodic table

The periodic table initially was based upon the chemical reactivity of the elements. Now it is a systematic arrangement of the elements based upon their atomic number.

Determining the Atomic Number of an Element Sample Exercise
   1. How many protons are present in one atom of the element titanium, Ti?
             The correct answer is 22.

Look on a periodic table for the element titanium, Ti. It is on the 4th row in the middle section of the periodic table. The square containing Ti looks like this.

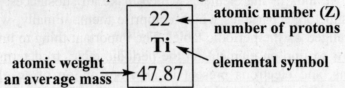

As indicated, the topmost number tells us the number of protons in the element. The bottom number, the atomic weight, is a weighted average of all the stable isotopes in the element.

Determining the Number of Electrons in an Element Sample Exercise
   2. How many electrons are present in one atom of the element titanium, Ti?
             The correct answer is 22.

For neutral atoms, atoms that have no net electrical charge, the number of electrons is equal to the number of protons.

Neutral **atoms** have electron numbers *equal* to the proton number.
Positive **ions** have *fewer* electrons than protons.
Negative **ions** have *more* electrons than protons.

Determining the Number of Neutrons in an Isotope Sample Exercise
   3. How many protons, neutrons, and electrons are present in one atom of the ^{50}Ti isotope?
             The correct answer is 22 protons, 28 neutrons, and 22 electrons.

Isotopes are elements having different numbers of neutrons in their nucleus. We write isotopes using nuclear symbols such as ^{50}Ti. Atoms of Ti have been found that have as few as 18 neutrons and as many as 32 neutrons.   Most of these Ti isotopes are radioactive (they are unstable leading to radioactive decomposition to other elements). However, five Ti isotopes – ^{46}Ti, ^{47}Ti, ^{48}Ti, ^{49}Ti, and ^{50}Ti – are stable. To determine the number of neutrons in an isotope subtract the proton number from the atomic mass.
             A - Z = N for this example 50 – 22 = 28 neutrons.

| atomic mass $A = Z + N$ | $\longrightarrow$ | ^{50}Ti | $\longleftarrow$ | Ti has an atomic number, Z, of 22 |

Determining the Number of Protons, Neutrons and Electrons in an Ion Sample Exercise

4. *How many protons, neutrons, and electrons are present in one atom of the $^{46}Ti^{2+}$ isotope?*

The correct answer is 22 protons, 24 neutrons, and 20 electrons.

Ti always has 22 protons. Other elements have different numbers of protons. To determine the number of neutrons for any problem, A - Z = N, which in this case gives 46 – 22 = 24 neutrons. The 2+ symbol is the ionic charge. A 2+ charge indicates that the atom has lost 2 electrons and now has fewer electrons than protons. Thus the number of electrons is 22 – 2 = 20.

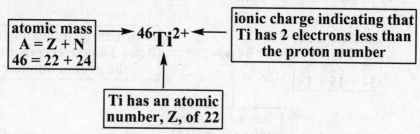

5. *How many protons, neutrons, and electrons are present in one atom of the $^{34}P^{3-}$ isotope?*

The correct answer is 15 protons, 19 neutrons, and 18 electrons.

Let's use our problem solving check list for this exercise.
1. **Read the problem**
    a) Pay special attention to the isotopic symbol $^{34}P^{3-}$. There is an enormous amount of information in this symbol.
2. **Understand precisely what the problem is asking**
    We must determine:
    a) The number of protons in a P atom.
    b) The number of neutrons in a ^{34}P isotope.
    c) The number of electrons in a P^{3-} ion.
3. **Identify the chemical principles**
    a) Comprehending atomic, isotopic, and ionic symbols.
    b) Finding proton numbers on the periodic table.
    c) Calculating neutron numbers from A = Z + N.
    d) Calculating the number of electrons from the ionic charge.
4. **Determine the relevant pieces of information**
    a) The most relevant piece of information presented in this problem is the isotopic symbol, $^{34}P^{3-}$.
    b) You must also know to look on the periodic table for the atomic number of P.
5. **Create a solution pathway**
    a) Determine the atomic number of P from the periodic table which is the bottom number in the box for P and is equal to 15.

21

b) Recognize that the superscript 34 is the atomic mass or A value for this isotope of P.

c) Since $A = Z + N$, then $N = A - Z$ giving $N = 34 - 15 = 19$.

d) The superscript 3- indicates that this is a negative ion which has 3 more electrons than protons.

e) The number of electrons is $15 + 3 = 18$.

6. Check your answer

a) The number of protons, 15, is correct for P because it appears in the P box on the periodic table.

b) 15 protons + 19 neutrons = 34 the atomic mass

c) 18 electrons = 15 protons + 3 electrons from the 3- ionic charge

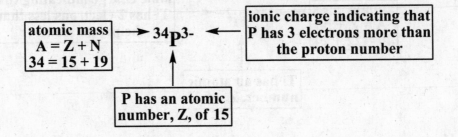

| atomic mass $A = Z + N$ $34 = 15 + 19$ | $\rightarrow$ | $^{34}P^{3-}$ | $\leftarrow$ | ionic charge indicating that P has 3 electrons more than the proton number |

P has an atomic number, Z, of 15

YIELD

1. Proton number, **Z**, is the top number in the element's square on the periodic table.

2. Atomic mass, **A**, is the left superscript number in the atom's nuclear symbol.

3. Neutron number, **N**, is **A - Z**.

4. Electron number is determined from the proton number and the ionic charge (right superscript number).
   a. For positive ions, the electron number = **Z** – ionic charge
   b. For negative ions the electron number = **Z** + ionic charge

INSIGHT:

You will find that you are frequently asked for the proton and electron numbers in your chemistry course. An element's chemistry is primarily determined by these two factors. Neutron numbers are important in understanding nuclear chemistry and radioactivity.

Be certain that you know how to determine the proton, neutron, and electron numbers for any element, isotope, or ion.

Atomic Weight of an Element Sample Exercises

6. *The stable Ti isotopes along with their natural distribution are given in the following table. Use this information to determine the atomic weight of Ti.*

Isotope	Mass (amu)	Natural Distribution (%)
^{46}Ti	45.95	8.00
^{47}Ti	46.95	7.30
^{48}Ti	47.95	73.8
^{49}Ti	48.95	5.50
^{50}Ti	49.95	5.40

The correct answer is 47.89 amu.

To calculate the weighted average for the atomic weight of Ti we multiply the mass of each isotope by its natural distribution percentage then add all five calculations. Each percentage must be divided by 100 to convert the percentage to decimal form.

$$? \text{ atomic weight} = \left(\underbrace{0.0800}_{\substack{8.00\,\% \text{ written} \\ \text{as a decimal}}} \times \underbrace{45.95}_{\substack{\text{actual mass} \\ \text{of } ^{46}\text{Ti}}} \right) + \left(\underbrace{0.0730}_{\substack{7.30\,\% \text{ written} \\ \text{as a decimal}}} \times \underbrace{46.95}_{\substack{\text{actual mass} \\ \text{of } ^{47}\text{Ti}}} \right) + \left(\underbrace{0.738}_{\substack{73.8\,\% \text{ written} \\ \text{as a decimal}}} \times \underbrace{47.95}_{\substack{\text{actual mass} \\ \text{of } ^{48}\text{Ti}}} \right) +$$

$$\left(\underbrace{0.0550}_{\substack{5.50\,\% \text{ written} \\ \text{as a decimal}}} \times \underbrace{48.95}_{\substack{\text{actual mass} \\ \text{of } ^{49}\text{Ti}}} \right) + \left(\underbrace{0.0540}_{\substack{5.40\,\% \text{ written} \\ \text{as a decimal}}} \times \underbrace{49.95}_{\substack{\text{actual mass} \\ \text{of } ^{50}\text{Ti}}} \right) \text{ amu}$$

Our calculated mass does not equal 47.87 amu due to round off errors in the masses.

$$= (3.68 + 3.43 + 35.39 + 2.69 + 2.70) \text{ amu}$$
$$= 47.89 \text{ amu}$$

INSIGHT: Units in this calculation are atomic mass units (amu), a unit designed to convert the very small masses of individual atoms into simpler numbers. You will not be expected to know the actual masses of the individual isotopes or natural distribution percentages. Frequently these values are given in the problem.

7. *The element boron, B, has two stable isotopes, ^{10}B and ^{11}B. The natural abundances for these two isotopes are 19.90% for ^{10}B and 80.10% for ^{11}B. The actual mass of ^{11}B is 11.01 amu. Determine the actual mass of ^{10}B.*

The correct answer is 10.0 amu.

This example is just a slight variation of the previous example. We are not given the mass of the ^{10}B isotope but we can readily determine it using the atomic weight of B from the periodic table, 10.81 amu, and the other information. All that is required is a little algebra as shown below.

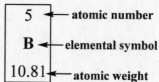

5 ← atomic number
B ← elemental symbol
10.81 ← atomic weight

23

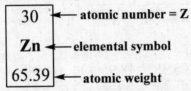

$$\text{atomic weight of B} \longrightarrow 10.81 \text{ amu} = \left(\underset{\substack{\text{19.9\% written} \\ \text{as a decimal}}}{0.1990} \; \underset{\substack{\text{unknown mass} \\ \text{of } ^{10}\text{B}}}{x} \right) + \left(\underset{\substack{\text{80.1\% written} \\ \text{as a decimal}}}{0.8010} \times \underset{\substack{\text{mass of } ^{11}\text{B}}}{11.01} \right) \text{amu}$$

$$10.81 \text{ amu} = (0.1990\,x) + 8.819 \text{ amu}$$

$$(10.81\text{-}8.819)\text{amu} = (0.199x)$$

$$1.99 \text{ amu} = (0.199x)$$

$$\frac{1.99 \text{ amu}}{0.199} = x$$

$$10.0 \text{ amu} = x$$

Periodic Table Sample Exercises

8. *What is the name and chemical symbol of the element that has Z = 30?*

    The correct answer is zinc, Zn.

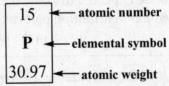

30 ← atomic number = Z

Zn ← elemental symbol

65.39 ← atomic weight

From earlier exercises we know that Z = 30 tells us that the proton number of this element is 30. Looking for element #30 we see that it is on the 4th row in the center portion of the periodic table. Its elemental symbol is Zn which stands for zinc.

9. *What is the name and chemical symbol of the element that has an atomic weight of 30.97 amu?*

    The correct answer is phosphorus, P.

15 ← atomic number

P ← elemental symbol

30.97 ← atomic weight

Only phosphorus, element #15, has an atomic weight of 30.97 amu.

YIELD

In this chemistry course you will frequently use the atomic numbers and atomic weights of the elements. Make sure you get the correct value from the periodic table. **Do not confuse the two numbers.**

Module 5
Chemical Nomenclature

Introduction

In this module, we examine how to name a variety of simple inorganic chemical compounds. The key to nomenclature is to <u>a) recognize the type of chemical compound then b) apply the appropriate nomenclature rules</u>. In this module, pay special attention to which elements are present in the compound and whether or not the elements are metals, transition metals, or nonmetals. Locations of the metals, transition metals, and nonmetals on the periodic table are shown in the schematic of the periodic table given after the Key Equations and Concepts.

Module 5 Key Equations & Concepts
Ionic Compounds

1. **Metal cations combined with nonmetal anions**.
   Simple binary ionic compounds are named using the metal's name followed by nonmetal's stem plus –ide.

2. **Metal cations combined with polyatomic anions**.
   Pseudobinary ionic compounds are named using the metal's name followed by polyatomic ion's name.

3. **Transition metal cations combined with nonmetal anions**.
   Transition metal ionic compounds are named using the metal's name with ionic charge in parentheses followed by nonmetal's stem plus –ide.

Covalent Compounds

4. **Two nonmetals combined in one compound**.
   In binary covalent compounds the less electronegative element is named first and the more electronegative named second using stem plus –ide.
   Prefixes such as di-, tri-, etc. are used for both elements.

5. **Hydrogen combined with a nonmetal in aqueous solution**.
   Binary acids are named using prefix hydro followed by nonmetal stem with suffix –ic acid.

6. **Hydrogen, oxygen, and a nonmetal combined in one compound**.
   Ternary acids are named based upon variations in the number of O atoms from the "-ic acid.".
   > One O atom more than the -ic acid is per stem –ic acid.
   > Look in your textbook for the formula of the –ic acid.
   > One O atom less than the -ic acid is stem –ous acid.
   > Two O atoms less than the -ic acid is hypo stem –ous acid.

Ionic Compounds Containing Ternary Acid Anions

7. **Metal ions combined with a polyatomic ternary acid anion**.
   Ternary acid salts are named using the metal's name followed by the same series of names used for the ternary acid with two changes. The –ic suffixes are changed to –ate and the –ous suffixes are changed to –ite.

Alkali metals

Metals

Nonmetal

Metalloi

Noble gases

1A
(1)

Alkaline earth metals

Transition metals

Halogens

8A
(18)

1	2A (2)											3A (13)	4A (14)	5A (15)	6A (16)	7A (17)	2 He
1 H																	
2	3 Li	4 Be										5 B	6 C	7 N	8 O	9 F	10 Ne
3	11 Na	12 Mg	3B (3)	4B (4)	5B (5)	6B (6)	7B (7)	8B (8) (9) (10)	1B (11)	2B (12)	13 Al	14 Si	15 P	16 S	17 Cl	18 Ar

Transition metals (3B – 2B):

4	19 K	20 Ca	21 Sc	22 Ti	23 V	24 Cr	25 Mn	26 Fe	27 Co	28 Ni	29 Cu	30 Zn	31 Ga	32 Ge	33 As	34 Se	35 Br	36 Kr
5	37 Rb	38 Sr	39 Y	40 Zr	41 Nb	42 Mo	43 Tc	44 Ru	45 Rh	46 Pd	47 Ag	48 Cd	49 In	50 Sn	51 Sb	52 Te	53 I	54 Xe
6	55 Cs	56 Ba	57 La *	72 Hf	73 Ta	74 W	75 Re	76 Os	77 Ir	78 Pt	79 Au	80 Hg	81 Tl	82 Pb	83 Bi	84 Po	85 At	86 Rn
7	87 Fr	88 Ra	89 Ac †	104 Rf	105 Db	106 Sg	107 Bh	108 Hs	109 Mt	110 Ds	111 Rg	112 (Uub)	--	114 (Uuq)	--	116 (Uuh)	--	--

*	58 Ce	59 Pr	60 Nd	61 Pm	62 Sm	63 Eu	64 Gd	65 Tb	66 Dy	67 Ho	68 Er	69 Tm	70 Yb	71 Lu
†	90 Th	91 Pa	92 U	93 Np	94 Pu	95 Am	96 Cm	97 Bk	98 Cf	99 Es	100 Fm	101 Md	102 No	103 Lr

Chemical Nomenclature Sample Exercises
Ionic Compounds

1. What is the correct name of this chemical compound, $CaBr_2$?

   The correct name is calcium bromide.

Ca^{2+} is a positive ion made from a metal.

Br^- ions are negative ions made from the nonmetal bromine.

Metal cations and nonmetal anions make simple binary ionic compounds. Simple binary ionic compounds are named using the metal's name followed by the nonmetal's stem and the suffix –ide. *No prefixes like di- or tri- are used to denote the number of ions present in the substance.*

1. What is the correct name of this chemical compound, $Mg_3(PO_4)_2$?

   The correct name is magnesium phosphate.

Mg^{2+} is a positive ion made from a metal.

PO_4^{3-} is a negative polyatomic ion made from phosphoric acid.

Metal cations and polyatomic anions make pseudobinary ionic compounds. These compounds are named using the metal's name followed by the correct name of the polyatomic anion. Polyatomic ions are positive or negative ions made from more than one atom. Below in Sample Exercise 7, the rules for naming polyatomic anions are given. Your textbook and instructor will indicate which of the polyatomic anions you are expected to know. Make sure that you have the name, the anion's formula, and the

26

charge memorized. ___Once again, no prefixes are used in these compounds to indicate the number of ions present.___

2. ___What is the correct name of this chemical compound, FeCl₃?___
   The correct name is iron(III) chloride.

This is a good example to apply our problem solving check list.

1. Read the problem
 a) The chemical symbol $FeCl_3$ is the key to solving this problem. A lot of information is contained in this symbol.

2. Understand precisely what the problem is asking
 a) We are being asked to name a compound made of a transition metal ion (Fe^{3+}) and a nonmetal ion (Cl^-).

3. Identify the chemical principles
 a) This is a nomenclature (naming) problem.
 b) The compound is composed of a metal and a nonmetal indicating it is an ionic compound.
 c) It is an ionic compound made of a transition metal ion (Fe^{3+}) and a nonmetal ion (Cl^-).

4. Determine the relevant pieces of information
 a) The most relevant piece of information is the compound formula $FeCl_3$.
 b) We know that Fe is a transition metal because it lies in the center section of the periodic table.
 c) We determine the iron ion's charge by knowing that Cl forms almost exclusively 1- ions. Because there are 3 Cl^- ions in $FeCl_3$ giving a charge of 3-, for the compound to be neutral the Fe ion must have a 3+ charge.

5. Create a solution pathway
 a) Since this is an ionic compound, we cannot use prefixes.
 b) Ionic compounds made of a transition metal and a nonmetal ion are named by placing the name of the metal ion first with its ionic charge inside parenthesis followed by the negative ion's stem plus the suffix –ide.
 c) The compound's name is iron(III) chloride .

6. Check your answer
 a) The compound is ionic.
 b) It is made of a transition metal ion and a nonmetal ion.
 c) The answer must not have prefixes.
 d) We must use the transition metal's name with charge in parenthesis and the nonmetal's stem plus –ide.
 e) Iron(III) chloride is the correct answer.

> ___Notice that in all of the compounds up to this point, prefixes have not been used to indicate the number of ions present.___

Covalent Compounds
3. ___What is the correct name of this chemical compound, N₂O₄?___
   The correct name is dinitrogen tetroxide.

This compound is made from two nonmetals, nitrogen and oxygen.

Two nonmetals form <u>binary covalent compounds</u>. ***Binary covalent compounds use prefixes to indicate the number of atoms of each element present in the compound.*** This is an important difference from the previous ionic compounds.

> 4. ***What is the correct name of this chemical compound, $H_2S(aq)$?***
>> The correct name is hydrosulfuric acid.
>> This compound is made from the nonmetal hydrogen and another nonmetal.

The symbol (aq) indicates that this compound is dissolved in water. The combination of hydrogen and a nonmetal dissolved in water is indicative of a <u>binary acid</u>. Binary acids are named using the prefix hydro- followed by the nonmetal's stem and the suffix –ide.

> 5. ***What is the correct name of this chemical compound, HIO_3?***
>> The correct name is iodic acid.
>> This compound is made from three nonmetals, H, O, and another nonmetal, iodine.

This combination of nonmetals is a <u>ternary acid</u>. Ternary acids are named based on a system derived from the third nonmetal's chemical state. The easiest method to learn how to name these compounds is to look in your textbook at the table of acids whose names end in "ic." You must learn both the compound's formula and its name. Once you know the acids that end in "ic", use the following system. The acid with one more O atom than the "ic acid" is the "per stem ic acid". One fewer O atom than the "ic acid" is the "ous acid". Two fewer O atoms than the "ic acid" is the "hypo stem ous acid". Here is the entire series for the iodine ternary acids.

HIO_4 is periodic acid. Periodic acid contains one more O atom than HIO_3. HIO_3 is iodic acid, the "ic" acid. HIO_2, with one less O atom than HIO_3, is iodous acid. Finally, HIO is hypoiodous acid. You are expected to know these various acids and the negative ions derived from them.

Ionic Compounds Containing Ternary Acid Anions
> 6. ***What is the correct name of this chemical compound, KIO_4?***
>> The correct name is potassium periodate.
>> This compound is made from a metal ion, K^+, and a polyatomic anion that is derived from the ternary acids discussed above.

This compound is a <u>ternary acid salt</u>. The anion's name is based upon the ending of the ternary acid. Ternary acids ending in "ic" give salts that end in "ate". Ternary acids that end in "ous" give salts that end in "ite". The prefixes per- and hypo- are retained. The iodic acid series of potassium salts are given below.

KIO_4 is potassium periodate. KIO_3 is potassium iodate. KIO_2 is potassium iodite. Finally, KIO is potassium hypoiodite.

	Ternary acid salts are very difficult compounds to name.
YIELD	Be sure to work hard on these.

Names of Some Common Chemical Species

Many chemical species are found in your home or at the grocery store. Their common names are used so frequently that your textbook or instructor may use the common name as well. For example, using the rules outlined above, H_2O should be named dihydrogenmonoxide. However, chemists and others refer to it as water. Similarly, NH_3 should be trihydrogenmonitride but we all know it as ammonia. Here are a few more chemical compounds whose common names you have heard and which will be discussed in this course.

Common Name	Correct Chemical Name	Chemical Formula
Baking Soda	Sodium hydrogen carbonate	$NaHCO_3$
Grain Alcohol	Ethanol	C_2H_5OH
Lime	Calcium oxide	CaO
Limestone	Calcium carbonate	$CaCO_3$
Milk of Magnesia	Magnesium hydroxide	$Mg(OH)_2$
Salt	Sodium chloride	$NaCl$
Sugar	Sucrose	$C_{12}H_{22}O_{11}$
Vinegar	Ethanoic or Acetic acid	$CH_3CO_2H(aq)$
Wood alcohol	Methanol	CH_3OH

Module 6
Understanding Chemical Formulas and Using the Mole Concept

Introduction

In this module we will look at how to interpret the information contained in a chemical formula. We will also examine the mole concept which provides a method to determine the numbers of atoms, ions, or molecules present in a measurable sample of any substance. We shall examine a) chemists' symbolic language, b) learn to recognize the number and types of atoms or ions present in a chemical compound, and c) how to perform the important calculations involving the mole concept. As you study these exercises you will need a periodic table for the atomic weights.

Module 6 Key Equations & Concepts

Chemical Formulas

1. C_5H_{12}

   Molecular formulas indicate the number of each atom present in a molecule.

2. $Al_2(CO_3)_3$

   Ionic formulas indicate the number of each ion present in a formula unit.

3. $4\ C_5H_{12}$

   Stoichiometric coefficients indicate the number, usually in moles, of a particular molecule or formula unit in the chemical symbolism.

Mole Concept

4. $\text{Molar mass} = \sum \text{atomic weights of atoms in a molecule or ion}$

   The molar mass, molecular weight, or formula weight equation is used to determine the mass in grams or amu of one mole of a substance.

5. $1\text{ mole} = 6.022 \times 10^{23}\text{ particles}$

   Avogadro's relationship is used to convert from the number of moles of a substance to the number of atoms, ions, or molecules of a substance and vice versa.

6. $\text{Mass of one atom of an element} = \left(\dfrac{\text{atomic mass of an element}}{1\text{ mole of an element}} \right)\left(\dfrac{1\text{ mole of an element}}{6.022 \times 10^{23}\text{ atoms}} \right)$

   This relationship is used to determine the mass of a few atoms, ions, or molecules of a substance.

7. $\text{Percentage composition} = \dfrac{\text{mass of one element in a compound}}{\text{molar mass of the compound}} \times 100\%$

   Percentage composition is used to experimentally determine the empirical and molecular formulas of compounds.

8. **Empirical formula**

   The simplest whole number ratio of elements in the compound

9. **Molecular formula**

   An integral multiple of the empirical formula indicating the actual numbers of atoms of each element in the compound

---

Interpreting Chemical Formulas Sample Exercises

1. How many atoms of each element are present in one molecule of C_2H_5OH?

The correct answer is 2 carbon atoms, 1 oxygen atom, and 6 hydrogen atoms.

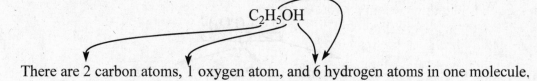

There are 2 carbon atoms, 1 oxygen atom, and 6 hydrogen atoms in one molecule.

Molecular formulas indicate the number of atoms
in each molecule of a covalent compound.

2. How many atoms of each element are present in one formula unit of $Al_2(SO_4)_3$?

The correct answer is 2 aluminum atoms, 3 sulfur atoms, and 12 oxygen atoms.

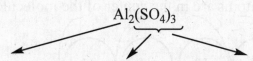

There are 2 aluminum atoms, 3 sulfur atoms, and 12 oxygen atoms.

Ionic compounds are written using formula units.
Just as for molecular formulas, formula units indicate the
number of atoms in each unit of an ionic compound.

**In all chemical formulas, numbers outside a parenthesis
are multiplied times the subscripts of the elements
inside the parenthesis.**

In $Al_2(SO_4)_3$ there are :

$2 \times 1 = 2$ aluminum atoms

$3 \times 1 = 3$ sulfur atoms

$3 \times 4 = 12$ oxygen atoms

Using Stoichiometric Coefficients Sample Exercises

*3. How many atoms of each element are present in this chemical formula
including the stoichiometric coefficient, $3\ C_5H_{12}$?*

The correct answer is 15 C atoms and 36 hydrogen atoms.

15 carbon atoms   36 hydrogen atoms

Stoichiometric coefficients are multiplied times
all of the atoms in the chemical compound.
$3 \times 5 = 15$ C atoms and $3 \times 12 = 36$ H atoms

4. How many atoms of each element are present in this chemical formula including the stoichiometric coefficient, $5\ Ca_3(PO_4)_2$?

The correct answer is 15 calcium atoms, 10 phosphorus atoms, and 40 oxygen atoms.

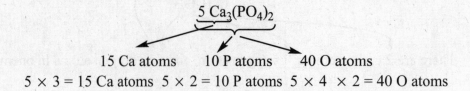

15 Ca atoms   10 P atoms   40 O atoms

$5 \times 3 = 15$ Ca atoms $5 \times 2 = 10$ P atoms $5 \times 4 \times 2 = 40$ O atoms

Interpreting Chemical Formulas Sample Exercises

5. Using circles to represent each atom draw a representation of the C_4H_{10} molecule.

$$C_4H_{10}$$

The 4 carbon atoms are in the center of the molecule.

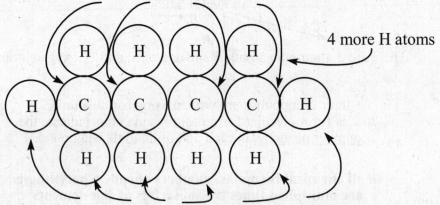

4 more H atoms

The 10 H atoms are around the outside of the molecule.
Notice, that this is one single molecule not 14 separate things.

It is possible to draw different molecules having the formula C_4H_{10} but all of the possibilities consist of molecules with atoms connected differently.

32

6. **Using circles to represent the atoms and ions, draw a representation of the ionic compound Al₂(SO₄)₃. Remember, one difference between ions and molecules is that ions are independent species while molecules are connected.**

The two aluminum ions are separate, independent species.

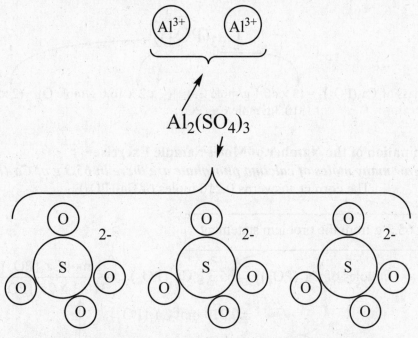

The three sulfate ions are also separate, independent species.

All of the formulas and symbols used so far can also represent moles of a species. If asked how many atoms, ions, or molecules are in one mole of a species, simply multiply the results from the previous exercises by Avogadro's number, 6.022×10^{23}. Remember that 1 mole $= 6.022 \times 10^{23}$, just like 1 dozen = 12. This next problem illustrates the idea.

Using Chemical Formulas to Determine Numbers of Atoms in One Mole of a Substance Sample Exercise

7. **How many atoms of each element are present in one mole of Al₂(SO₄)₃?**

$$Al_2(SO_4)_3$$

There are:

$2 \times 6.022 \times 10^{23} = 12.04 \times 10^{23}$ aluminum atoms

$3 \times 6.022 \times 10^{23} = 18.07 \times 10^{23}$ sulfur atoms

$12 \times 6.022 \times 10^{23} = 72.26 \times 10^{23}$ oxygen atoms

33

Determination of the Molar Mass Sample Exercise

   8. What is the molar mass or formula weight of calcium phosphate, $Ca_3(PO_4)_2$?
   The correct answer is 310.3 g/mol of $Ca_3(PO_4)_2$.

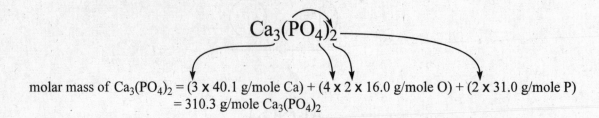

molar mass of $Ca_3(PO_4)_2$ = (3 x 40.1 g/mole Ca) + (4 x 2 x 16.0 g/mole O) + (2 x 31.0 g/mole P)
              = 310.3 g/mole $Ca_3(PO_4)_2$

Determination of the Number of Moles Sample Exercise

   9. How many moles of calcium phosphate are there in 65.3 g of $Ca_3(PO_4)_2$?
   The correct answer is 0.210 moles of $Ca_3(PO_4)_2$.

> 65.3 g from the problem statement

$$? \text{ moles of } Ca_3(PO_4)_2 = 65.3 \text{ g } Ca_3(PO_4)_2 \left(\frac{1 \text{ mol } Ca_3(PO_4)_2}{310.3 \text{ g } Ca_3(PO_4)_2} \right)$$

$$= 0.210 \text{ mol } Ca_3(PO_4)_2$$

> molar mass of calcium phosphate from exercise #8.

Once we know the number of moles of a substance, we can determine the number of molecules or formula units of the substance. (Molecules are found in covalent compounds. Ionic compounds do not have molecules, thus their smallest subunits are named formula units.) The key to solving these problems is properly using Avogadro's number.

Determination of the Number of Molecules or Formula Units Sample Exercise

   10. How many formula units of calcium phosphate are there in 0.210 moles of $Ca_3(PO_4)_2$?
   The correct answer is 1.26×10^{23} formula units of $Ca_3(PO_4)_2$.

0.210 moles from the problem statement	Avogadro's number converting moles to formula units

$$? \text{ formula units of } Ca_3(PO_4)_2 = 0.210 \text{ moles of } Ca_3(PO_4)_2 \left(\frac{6.022 \times 10^{23} \text{ formula units}}{1 \text{ mole of } Ca_3(PO_4)_2} \right)$$

$$= 1.26 \times 10^{23} \text{ formula units of } Ca_3(PO_4)_2$$

Determination of the Number of Atoms or Ions Sample Exercise

11. How many oxygen, O, atoms are there in 0.210 moles of $Ca_3(PO_4)_2$?

The correct answer is 1.02×10^{24} oxygen atoms.

From exercise 10

In $Ca_3(PO_4)_2$ there are $4 \times 2 = 8$ oxygen atoms.

$$? \, O \, atoms = 1.27 \times 10^{23} \text{ formula units of } Ca_3(PO_4)_2 \left(\frac{8 \text{ oxygen atoms}}{1 \text{ formula unit of } Ca_3(PO_4)_2} \right)$$

$$= 1.02 \times 10^{24} \text{ oxygen atoms}$$

Exercise 12 uses equation 6 from the Key Equations and Concepts to determine the mass of a few molecules or formula units of two compounds.

Determination of the Mass of Small Numbers of Molecules or Formula Units of a Substance Sample Exercise

12. What is the mass, in grams, of 25.0 formula units of $Ca_3(PO_4)_2$?

The correct answer is 1.29×10^{-20} g.

Avogadro's relationship

Molar mass of $Ca_3(PO_4)_2$ as a unit factor.

$$? \, g = 25.0 \text{ formula units of } Ca_3(PO_4)_2 \left(\frac{1 \text{ mole}}{6.022 \times 10^{23} \text{ formula units}} \right) \left(\frac{310.3 \text{ g of } Ca_3(PO_4)_2}{1 \text{ mole of } Ca_3(PO_4)_2} \right)$$

$$= 1.29 \times 10^{-20} \text{ g}$$

Combined Equations Sample Exercise

13. How many carbon, C, atoms are there in 0.375 g of $C_4H_8O_2$?

The correct answer is 1.03×10^{22} C atoms.

This is a good problem to show how the problem solving check list is relevant in solving such a relatively complicated problem.

1. **Read the problem**
2. **Understand precisely what the problem is asking**
   a) This problem is asking us to calculate the number of C atoms in 0.375 g of the compound $C_4H_8O_2$.
3. **Identify the chemical principles**
   a) This problem is a combination of the mole concept, understanding chemical formulas, and using Avogadro's number.
4. **Determine the relevant pieces of information**
   a) The relevant pieces of information are the mass of the compound (0.375 g) and the chemical formula ($C_4H_8O_2$).

b) We must also know the mole concept and how to use Avogadro's number (6.022×10^{23}) correctly.

5. Create a solution pathway

a) A good pathway for this problem is:

$$mass \rightarrow moles \rightarrow number\ of\ molecules \rightarrow number\ of\ atoms$$

as shown worked out below

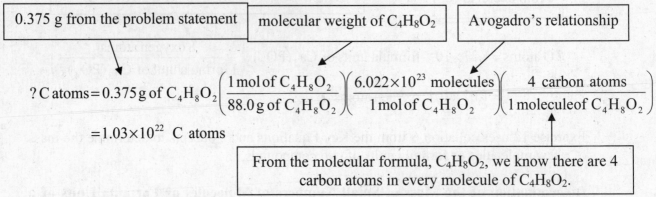

| 0.375 g from the problem statement | molecular weight of $C_4H_8O_2$ | Avogadro's relationship |

$$? \text{C atoms} = 0.375\,\text{g of } C_4H_8O_2 \left(\frac{1\,\text{mol of } C_4H_8O_2}{88.0\,\text{g of } C_4H_8O_2} \right) \left(\frac{6.022 \times 10^{23}\,\text{molecules}}{1\,\text{mol of } C_4H_8O_2} \right) \left(\frac{4\,\text{carbon atoms}}{1\,\text{molecule of } C_4H_8O_2} \right)$$

$$= 1.03 \times 10^{22}\ \text{C atoms}$$

From the molecular formula, $C_4H_8O_2$, we know there are 4 carbon atoms in every molecule of $C_4H_8O_2$.

6. Check your answer

a) This sample of $C_4H_8O_2$ is a visible amount of $C_4H_8O_2$. Because we can see it, there must be incredibly large numbers of molecules of $C_4H_8O_2$ and C atoms.

b) Our answer of 1.03×10^{22} C atoms is reasonable, has the correct units, and therefore is correct.

Percentage Composition Sample Exercise

14. Determine the percentage composition of the elements in sucrose, $C_{12}H_{22}O_{11}$. The molar mass of sucrose is 342.0 g/mol.

The correct answer is 42.1% C, 6.4% H, and 51.5% O.

| atom numbers from molecular formula | atomic masses from periodic table |

$$\% \text{ composition C} = \frac{12 \text{ atoms C} \times 12.0\ \text{amu}/\text{C atom}}{342.0\ \text{amu}} \times 100\% = 42.1\%\ C$$

$$\% \text{ composition H} = \frac{22 \text{ atoms H} \times 1.0\ \text{amu}/\text{H atom}}{342.0\ \text{amu}} \times 100\% = 6.4\%\ H$$

$$\% \text{ composition O} = \frac{11 \text{ atoms O} \times 16.0\ \text{amu}/\text{O atom}}{342.0\ \text{amu}} \times 100\% = 51.5\%\ O$$

Empirical Formula Sample Exercise

15. A compound composed entirely of nitrogen and oxygen has a percentage composition of 30.4% nitrogen and 69.6% oxygen. Determine its empirical formula.

The correct answer is NO_2.

The percentage composition is correct for any amount of this compound. For example, in a 5.00 g sample, 30.4% of the sample is N atoms and 69.6% is O atoms. Knowing this, let's choose an amount that will make determining the mass of the N and O in the sample the easiest. That amount is 100.0 g because it is easy to take 30.4% and 69.6% of 100.0 g. We immediately know that there are 30.4 g of N and 69.6 g of O in a 100.0 g sample. This logic is correct for any empirical formula problem.

1) Determine the number of moles of each element from the amounts of each element.

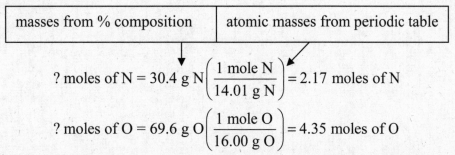

$$? \text{ moles of N} = 30.4 \text{ g N}\left(\frac{1 \text{ mole N}}{14.01 \text{ g N}}\right) = 2.17 \text{ moles of N}$$

$$? \text{ moles of O} = 69.6 \text{ g O}\left(\frac{1 \text{ mole O}}{16.00 \text{ g O}}\right) = 4.35 \text{ moles of O}$$

2) To determine the smallest whole number ratio of the elements, divide each molar amount by the smallest molar amount of any one element in the compound. For this compound 2.17 moles < 4.35 moles, so we divide both amounts by 2.17.

amounts from previous step

$$\frac{2.17 \text{ moles of N}}{2.17} = 1 \text{ N}$$

$$\frac{4.35 \text{ moles of O}}{2.17} = 2 \text{ O}$$

3) Step 2 gives us the simplest ratio of the number of atoms of each element in the compound which is the empirical formula, NO_2. The empirical formula tells us that in this compound there are 2 oxygen atoms for every 1 N atom. However, the empirical formula is the same for N_2O_4, N_3O_6, etc. In exercise 16, we will determine this compound's actual molecular formula from the empirical formula.

Molecular Formula Sample Exercise

16. *The compound used in exercise 15 has a molar mass of 92.0 g/mol. What is its molecular formula?*

The correct answer is N_2O_4.

The molecular formula must be a whole number multiple of the empirical formula. To find this whole number, divide the compound's molar mass by the mass of the empirical formula.

$$? \text{ mass of empirical formula} = \left(1 \times 14.01 \text{ g/mol N}\right) + \left(2 \times 16.00 \text{ g/mol O}\right) = 46.01 \text{ g/mol}$$

$$? \text{ whole number ratio} = \frac{92.0 \text{ g/mol}}{46.01 \text{ g/mol}} = 2.0$$

molar mass from problem

empirical mass from previous step

The whole number ratio indicates that the molecular formula is the empirical formula doubled. In other words, this compound has 2 x 1 = 2 N atoms and 2 x 2 = 4 O atoms which gives the correct molecular formula N_2O_4.

Module 7
Understanding Chemical Reaction Equations

Introduction

This module is designed to help you understand and interpret chemical reactions as well as recognize some types of reactions. The important points to take from this module are a) how to interpret the information presented in chemical equations, b) how to balance chemical equations, and c) how to classify reaction types.

Module 7 Key Equations & Concepts

1. **Chemical equations**
   Equations are symbolic representations of the chemical changes that occur in a chemical reaction at the atomic/molecular level.

2. **Balancing chemical equations**
   A process using stoichiometric coefficients in the reaction equation to show that equal numbers of atoms of each element are present in the reactant and product sides of the equation. Balancing is necessary to obey the Law of Conservation of Mass.

3. **Combination reactions**
   Two or more chemical species react to form a new chemical species.

4. **Decomposition reactions**
   A single chemical species breaks apart forming two or more different chemical species.

5. **Single-replacement reactions**
   One chemical species displaces a chemical species in a more complex substance.

6. **Double-replacement reactions**
   Two chemical species displace chemical species in a chemical reaction.

Interpreting Chemical Reaction Sample Exercises

The symbols that chemists use to depict chemical reactions are designed to specify what the atoms are doing when a reaction occurs. For example, which atoms or ions are breaking chemical bonds then forming new bonds? Studying chemistry is much easier once you learn to visualize what reaction equations are indicating symbolically.

Some terminology required for this module is shown in the following diagram.

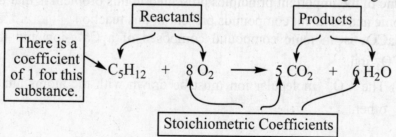

39

Stoichiometric coefficients are necessary to balance chemical equations. Balancing insures that equal numbers of atoms of each element are present in both the products and reactants. Otherwise the reaction would violate the Law of Conservation of Mass indicating that new atoms have been made in the reaction rather than redistributing the atoms into new chemical compounds.

Interpreting Combination Reaction Equations

1. *Using circles to represent the atoms, ions, or molecules, depict the following chemical reaction.*

$$2\,CO + O_2 \rightarrow 2\,CO_2$$

The correct answer is:

2 CO O_2 $2\,CO_2$

> **INSIGHT:** Subscripted numbers indicate the number of atoms or ions in an ionic or molecular compound. Numbers on the line, called stoichiometric coefficients, indicate the number of each ionic or molecular compound involved in the reaction. For example, 2 CO_2 tells us that there are 2 molecules of carbon dioxide and that each molecule is made of 1 C atom and 2 O atoms.

Interpreting Decomposition Reaction Equations

2. *Using circles to represent the atoms, ions, or molecules, depict the following chemical reaction.*

$$CaCO_3 \rightarrow CaO + CO_2$$

Let's study this reaction in more detail using the problem steps check list from Module 1.

1. **Read the problem**
   a) Pay careful attention to the chemical reaction while asking yourself:
      1) How will the atoms rearrange themselves in this reaction?
      2) How can we depict that rearrangement?
2. **Understand precisely what the problem is asking**
   a) This problem is asking us to visualize on the atomic/molecular level how this decomposition reaction occurs.
3. **Identify the chemical principles**
   a) One of the important principles presented in this problem is that there are both ionic and covalent compounds present in this reaction.
   b) $CaCO_3$ is an ionic compound composed of a Ca^{2+} ion and the molecular CO_3^{2-} ion.
      1) The CO_3^{2-} molecular ion must be drawn with the atoms attached to each other.

2) Also, the CO_2 molecule must be drawn with the atoms attached to each other.

c) CaO is an ionic compound. The atoms in CaO must not be attached to each other.

4. **Determine the relevant pieces of information**
   a) The most relevant piece of information in this problem is the chemical reaction and the compounds present in that reaction.
   b) There are no irrelevant pieces of information in this problem.

5. **Create a solution pathway**
   a) Draw each compound with the correct number of atoms of each element, with the correct ionic charges, and with the atoms attached or not attached properly as indicated in the previous steps.
   b) The correct answer is given below.

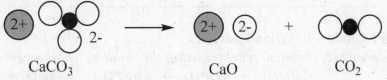

   CaCO$_3$ CaO   CO$_2$

6. **Check your answer**
   a) The reaction is correctly drawn with the number of atoms of each element properly balanced.
   b) Each compound is drawn with the appropriate atom-to-atom connections for molecules and with ionic compounds having separated atoms.
   c) Most importantly, in this reaction the carbonate ion, CO_3^{2-}, from $CaCO_3$, decomposes into one molecular compound carbon dioxide, CO_2, leaving the ionic oxide ion, O^{2-}, remaining in the compound calcium oxide, CaO.

Interpreting Single-replacement Equations

3. *Using circles to represent the atoms, ions, or molecules, depict the following chemical reaction.*

$$Fe + Ni(NO_3)_2 \rightarrow Fe(NO_3)_2 + Ni$$

The correct answer is:

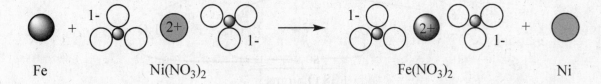

Fe   Ni(NO$_3$)$_2$ Fe(NO$_3$)$_2$ Ni

Notice that in this reaction, the nitrate ion, NO_3^-, is undisturbed but that the Fe atom replaces the Ni atom.

Interpreting Double-replacement Equations

4. *Using circles to represent the atoms, ions, or molecules, depict the following chemical reaction.*

$$AgNO_3 + NaCl \rightarrow AgCl + NaNO_3$$

The correct answer is:

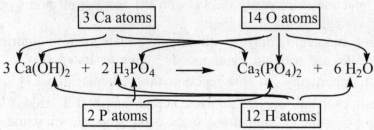

AgNO₃   NaCl   AgCl   NaNO₃

Notice that in this reaction, the silver, Ag^+, and sodium, Na^+, ions switch negative ions (nitrate ion, NO_3^-, and chloride ion, Cl^-) in the course of the reaction.

INSIGHT:	Look back at the four reactions depicted above. Work hard at understanding how the chemical reaction symbols show the atoms, ions, or molecules rearranging themselves to form the products. This skill will pay many dividends for you as the course proceeds.

Balancing Chemical Reaction Exercises

5. *Balance this chemical reaction using the smallest whole numbers.*

$$Ca(OH)_2 + H_3PO_4 \rightarrow Ca_3(PO_4)_2 + H_2O$$

    The correct answer is:

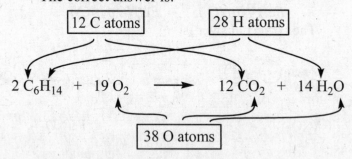

$$3\ Ca(OH)_2 + 2\ H_3PO_4 \longrightarrow Ca_3(PO_4)_2 + 6\ H_2O$$

6. *Balance this chemical reaction using the smallest whole numbers.*

$$C_6H_{14} + O_2 \rightarrow CO_2 + H_2O$$

    The correct answer is:

$$2\ C_6H_{14} + 19\ O_2 \longrightarrow 12\ CO_2 + 14\ H_2O$$

INSIGHT:	In most, but not all reactions, it is easiest to balance the chemical equation if you begin with the elements that are present in the fewest numbers. For example, in exercise 5 balance the Ca and P atoms first then tackle the O and H atoms. Once the Ca and P atoms are balanced the O and H atoms balance readily.

42

Combination Reaction Sample Exercise

7. ***What common reaction type is represented by this chemical reaction?***

$$3\, Sr\ +\ N_2\ \rightarrow\ Sr_3N_2$$

The correct answer is a combination reaction.

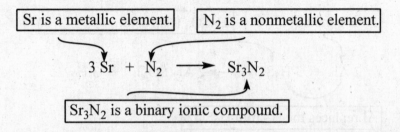

| Sr is a metallic element. | | N_2 is a nonmetallic element. |

$$3\, Sr\ +\ N_2\ \longrightarrow\ Sr_3N_2$$

Sr_3N_2 is a binary ionic compound.

INSIGHT: Combination reactions are characterized by *a) the reaction of two elements to form a compound, b) the reaction of a compound and an element to form a new compound, or c) the reaction of two compounds to form a new compound.*

Combination reactions, like most other reactions, may be classified as other reaction types. For example, the reaction of Sr with N_2 is also a reduction-oxidation (redox) reaction. We will address redox reactions later in this Survival Guide.

Decomposition Reaction Sample Exercise

8. ***What common reaction type is represented by this chemical reaction?***

$$2\, CaO\ \rightarrow\ 2\, Ca\ +\ O_2$$

The correct answer is decomposition reaction.

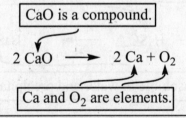

CaO is a compound.

$$2\, CaO\ \longrightarrow\ 2\, Ca + O_2$$

Ca and O_2 are elements.

INSIGHT: There are three types of decomposition reactions: *a) compounds decomposing into two or more elements, b) compounds decomposing into another compound and an element (the CaCO₃ reaction from Exercise 2 is an example), and c) compounds decomposing into two simpler compounds.*

Decomposition reactions are the reverse of combination reactions. Instead of putting elements or compounds together to make new compounds, decomposition reactions break compounds into elements or less complex compounds. As in combination reactions decomposition reactions can also be classified as more than one reaction type. This CaO decomposition reaction is also a redox reaction.

Single-replacement Reaction Sample Exercise

9. ***What common reaction type is represented by this reaction?***

$$2\ Al\ +\ 3\ H_2SO_4\ \rightarrow\ Al_2(SO_4)_3\ +\ 3\ H_2$$

The correct answer is single-replacement reaction.

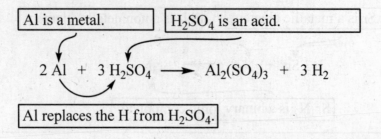

Al is a metal.	H_2SO_4 is an acid.

$$2\ Al\ +\ 3\ H_2SO_4\ \longrightarrow\ Al_2(SO_4)_3\ +\ 3\ H_2$$

Al replaces the H from H_2SO_4.

INSIGHT: Single-replacement reactions are characterized by *one element replacing a second element in a compound.* The three types of single-replacement reactions are: *a) an active metal displacing another metal from certain ionic compounds, b) an active metal displacing hydrogen from either HCl or H_2SO_4, and c) an active nonmetal displacing another nonmetal from certain ionic compounds.*

Double-replacement Reaction Sample Exercise

10. ***What common reaction type is represented by this reaction?***

$$Ba(OH)_2\ +\ H_2SO_4\ \rightarrow\ BaSO_4\ +\ 2\ H_2O$$

The correct answer is a double-replacement reaction.

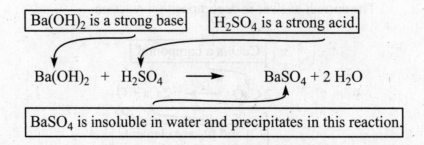

$Ba(OH)_2$ is a strong base. H_2SO_4 is a strong acid.

$$Ba(OH)_2\ +\ H_2SO_4\ \longrightarrow\ BaSO_4 + 2\ H_2O$$

$BaSO_4$ is insoluble in water and precipitates in this reaction.

INSIGHT: In metathesis reactions *the ions switch partners.*

This reaction is illustrated using the symbols AB to represent $Ba(OH)_2$ and CD to represent H_2SO_4. Products $BaSO_4$ and H_2O are represented using AD and CB respectively.

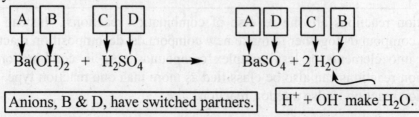

$$\boxed{A}\ \boxed{B}\quad \boxed{C}\ \boxed{D}\qquad\qquad \boxed{A}\ \boxed{D}\quad \boxed{C}\ \boxed{B}$$

$$Ba(OH)_2\ +\ H_2SO_4\ \longrightarrow\ BaSO_4 + 2\ H_2O$$

Anions, B & D, have switched partners. $H^+ + OH^-$ make H_2O.

When an acid reacts with a base, a salt is formed, $BaSO_4$ in this case, and water if the base is a hydroxide. Water is formed by the combination of the H^+ with the OH^-.

This reaction is also depicted below using circles to represent the atoms, ions, and molecules. Notice that the H atoms originally on H_2SO_4 migrate to the OH^- ions to make water as the Ba^{2+} ion combines with the resulting SO_4^{2-} ion to form $BaSO_4$.

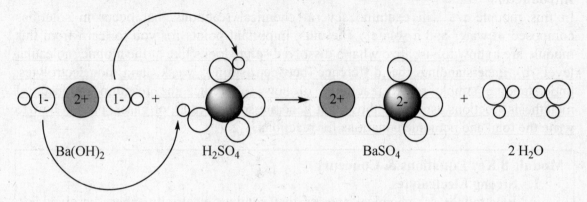

$Ba(OH)_2$ H_2SO_4 $BaSO_4$ $2\ H_2O$

Visualizing chemical reactions at the atomic level is a skill that can only be acquired through practice. For practice, draw circle representations of the chemical reactions presented in exercises 7, 8, and 9.

Module 8
Chemical Reactions in Aqueous Solutions

Introduction

In this module we will examine several chemical reactions that occur in solutions composed of water and a solute. The most important points for you to learn from this module are a) how to visualize what a dissolved solute looks like at the atomic/molecular level, b) understanding the difference between strong, weak, and nonelectrolytes, understanding simple redox reactions, c) how to predict the reaction products of metathesis reactions especially those that generate solids, water, or gases, plus d) how to write the total and net ionic equations for reactions.

Module 8 Key Equations & Concepts

1. **Strong Electrolytes**

   Electrolytes are chemical species that conduct electricity when dissolved in solution. Strong electrolytes conduct electricity in water especially well because 100% of the molecules ionize (split up into ions) or formula units dissociate (break apart) in solution.

2. **Weak Electrolytes**

   Weak electrolytes conduct electricity poorly in solution. For weak electrolytes much less than 100% (typically 10% or less) of the molecules or formula units dissociate or ionize in aqueous solution.

3. **Nonelectrolytes**

   Nonelectrolytes do not conduct electricity in solution because nonelectrolytes do not dissociate in water even though they do dissolve in solution.

4. **Reduction-Oxidation Reactions**

   Reduction-oxidation (redox) reactions occur when species exchange electrons. Oxidation occurs when a chemical substance loses electrons. Reduction occurs when a chemical substance gains electrons. To understand redox reactions we must determine the ionic charges of the species before and after the reaction occurs.

5. **$KOH(aq) + HI(aq) \rightarrow KI(aq) + H_2O(l)$ – Formula Unit Equation**

   The formula unit equation shows all of the species involved in a reaction as ionic or molecular compounds.

6. **$K^+(aq) + OH^-(aq) + H^+(aq) + I^-(aq) \rightarrow K^+(aq) + I^-(aq) + H_2O(l)$ – Total Ionic Equation**

   The total ionic equation shows all of the ions in their ionized states in solution. All species that ionize completely in water are shown as separated ions.

7. **$OH^-(aq) + H^+(aq) \rightarrow H_2O(l)$ – Net Ionic Equation**

   To write the net ionic equation, remove all spectator ions from the total ionic equation. Spectator ions are species that do not change as the reaction proceeds from reactants to products.

Strong Electrolyte, Weak Electrolyte, Nonelectrolyte Sample Exercises

1. *Shown below are three images of aqueous (water based) solutions drawn from the ionic or molecular perspective. Which image is the best representation of a strong electrolyte, a weak electrolyte, and a nonelectrolyte?*
   The correct answer is image A is a weak electrolyte, image B is a nonelectrolyte, and image C is a strong electrolyte.

Image Key

= H_2O = molecule   = + ion   = - ion

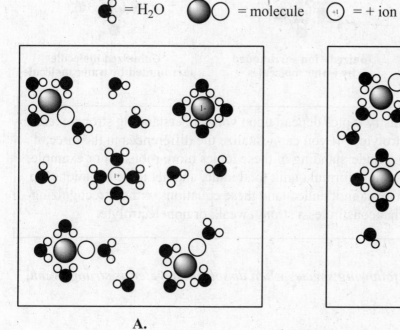

A.

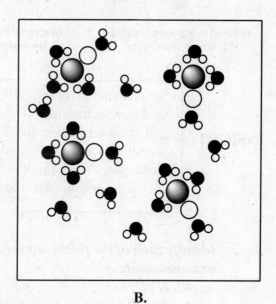

B.

C.

The extent that molecules ionize (break apart into ions) or ionic compounds dissociate (separate) in water or other solvents is the key to understanding electrolytes. Strong electrolytes ionize or dissociate ~ 100% in water. Weak electrolytes ionize or dissociate much less than 100%. Nonelectrolytes do not ionize or dissociate at all. In the images for this exercise notice that in image C all of the molecules have ionized yielding only ions dissolved in solution. Image A has 1 ionized molecule and 3 unionized molecules indicating this is an image of a weak electrolyte. There are no ionized molecules in image B. Every molecule in this image is surrounded by water molecules indicating a dissolved nonelectrolyte. Shown below are images of a dissolved positive ion, a dissolved negative ion, and a dissolved molecule.

Ionized + ion surrounded   **Ionized - ion surrounded**   **Unionized molecule**
by water molecules   **by water molecules**   **surrounded by water molecules**

INSIGHT: Several chemistry topics depend upon your understanding strong, weak, and nonelectrolytes. If you can visualize the differences in the three, it will make your understanding of these topics more robust. For example, later in this module formula unit, total ionic, and net ionic equations are discussed. You cannot understand these equations without recognizing what constitutes a strong, weak, or nonelectrolyte.

2. *Identify each of the following species when dissolved in water as strong, weak, or nonelectrolytes.*
   a. *HNO_3*
   b. *$Ca_3(PO_4)_2$*
   c. *CH_3CO_2H*
   d. *C_2H_5OH*
   e. *Na_2CO_3*

   The correct answer is a and e are strong electrolytes, c is a weak electrolyte, b and d are nonelectrolytes.

In your textbook there are sets of rules to help you recognize which compounds are soluble in water and which ionize/dissociate extensively. You will need to know these rules by heart. In exercise 2 HNO_3 is a strong, water soluble acid. All strong, water soluble acids are strong electrolytes. Na_2CO_3 is a strong, water soluble ionic compound. All strong, water soluble ionic compounds dissociate 100% in water. CH_3CO_2H is acetic or ethanoic acid. Acetic acid is a weak, water soluble covalent compound. Weak, water soluble compounds ionize much less than 100%. (In typical solution concentrations acetic acid ionizes less than 10%.) Calcium phosphate, $Ca_3(PO_4)_2$, is an ionic compound that is essentially insoluble in water. Insoluble chemical compounds do not ionize or dissociate in water therefore are nonelectrolytes. Ethanol, C_2H_5OH, is water soluble but it does not ionize in water. All compounds that dissolve in water but do not ionize are nonelectrolytes.

Reduction Oxidation Reaction Sample Exercise

3. *What reaction types are represented by this chemical reaction?*

$$2\ Na(s)\ +\ 2\ H_2O(l)\ \rightarrow\ 2\ NaOH(aq)\ +\ H_2(g)$$

The correct answer is reduction oxidation (redox) and combination reaction.

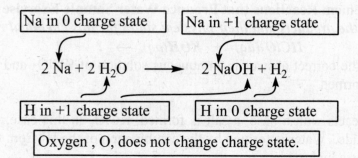

INSIGHT:	To recognize redox reactions you must look for chemical species that change their charge states.

Rules for assigning charge states were discussed in Module 5. Review those rules if you do not know them.

	Double replacement (or metathesis) reactions were described in Module 7. Refresh your memory before proceeding to the next exercises.

Double Replacement Precipitation Reactions Sample Exercise

4. *Identify the products and their physical states for this chemical reaction.*

$$Fe(NO_3)_2(aq)\ +\ (NH_4)_2CO_3(aq)\ \rightarrow\ ?\ +\ ?$$

The correct answer is solid $FeCO_3$ is formed and precipitates from solution while NH_4NO_3 remains dissolved in solution.

In double replacement reactions the ions switch partners forming two new products. For this reaction, the ion switching pattern is indicated below.

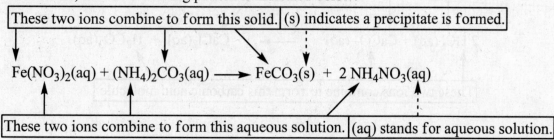

INSIGHT:	Precipitation reactions are characterized by ***the formation of a compound that is insoluble in water.***

YIELD *To determine if a solid forms and precipitates from solution you must know the solubility guidelines from your textbook.* We know that $FeCO_3$ precipitates because the solubility guidelines indicate that carbonate salts (except NH_4^+ and the alkali metal carbonates) are insoluble in water.

Double Replacement Reactions that Produce Water Sample Exercise

5. ***Identify the products and their physical states for this chemical reaction.***

$$HClO_3(aq) \;+\; KOH(aq) \;\rightarrow\; ? \;+\; ?$$

The correct answer is an aqueous solution of $KClO_3$ and liquid water are formed.

When an acid reacts with a base, a salt is formed, $KClO_3$ in this case, and water if the base is a hydroxide. Water is formed by the combination of the H^+ from the acid with the OH^- from the basic hydroxide.

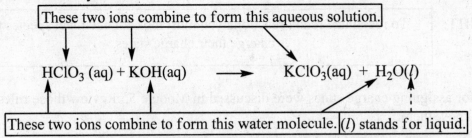

These two ions combine to form this aqueous solution.

$$HClO_3\,(aq) + KOH(aq) \longrightarrow KClO_3(aq) + H_2O(l)$$

These two ions combine to form this water molecule.	(l) stands for liquid.

INSIGHT: Acid-base reactions involving a basic hydroxide will **form a salt and water.**

Double Replacement Reactions that Produce Gases Sample Exercise

6. ***Identify the products and their physical states for this chemical reaction.***

$$HCl(aq) \;+\; CaCO_3(s) \;\rightarrow\; ? \;+\; ?$$

The correct answer is an aqueous solution of calcium chloride, liquid water, and gaseous carbon dioxide.

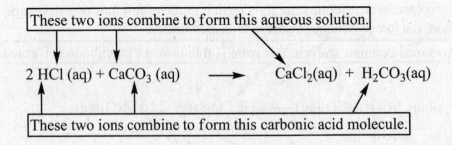

These two ions combine to form this aqueous solution.

$$2\,HCl\,(aq) + CaCO_3\,(aq) \longrightarrow CaCl_2(aq) + H_2CO_3(aq)$$

These two ions combine to form this carbonic acid molecule.

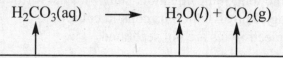

Carbonic acid is an unstable molecule.
It rapidly decomposes into two stable molecules.

$$H_2CO_3(aq) \longrightarrow H_2O(l) + CO_2(g)$$

Carbonic acid breaks apart into water and carbon dioxide.

INSIGHT: Double replacement reactions that produce gases behave exactly the same as the other metathesis reactions in the first reaction step, they switch ions. They differ from the other reactions because an unstable molecule is formed in the first step. The unstable molecule decomposes into water and a gas. ***Look for double displacement reactions that form carbonic acid (H_2CO_3) or sulfurous acid (H_2SO_3) in the first reaction step.*** Sulfurous acid decomposes into $H_2O(l)$ and $SO_2(g)$.

Total and Net Ionic Equations Sample Exercises

7. *Write the total ionic and net ionic equations for this reaction.*

$$Ba(OH)_2(aq) \ + \ 2\,HCl(aq) \ \rightarrow \ BaCl_2(aq) \ + \ 2\,H_2O(l)$$

The correct total ionic equation is:

$$Ba^{2+}(aq) + 2\,OH^-(aq) + 2\,H^+(aq) + 2\,Cl^-(aq) \rightarrow Ba^{2+}(aq) + 2\,Cl^-(aq) + 2\,H_2O(l)$$

The correct net ionic equation is:

$$2\,OH^-(aq) + 2\,H^+(aq) \rightarrow 2\,H_2O(l)$$

or

$$OH^-(aq) + H^+(aq) \rightarrow H_2O(l)$$

Unless you are readily familiar with the solubility guidelines, these problems are exceedingly difficult. This is a good exercise to use our problem solving steps check list. It will help to clarify the answer to this exercise.

1. **Read the problem**
   a) The formula unit equation is the reaction equation written just below the question for this exercise. It is the most important piece of information in this exercise.
2. **Understand precisely what the problem is asking**
   a) We are being asked to break up the formula unit equation into the total and net ionic equations for this reaction.
3. **Identify the chemical principles**
   a) We must use formula unit, total ionic, and net ionic equations.
   b) We must understand what spectator ions are.
   c) We must use the solubility guidelines.
   d) We must understand how the compounds in this reaction are composed of ions or molecules.
4. **Determine the relevant pieces of information**
   a) The formula unit equation is the most relevant piece of information.

b) We must also know the solubility guidelines and the ionic/molecular structure of the species involved in the reaction.

5. Create a solution pathway

a) To determine the total ionic equation, start with the formula unit equation and split each strong electrolyte into its constituent ions.

b) Weak and nonelectrolytes are not separated.

YIELD

In total and net ionic equations there are three classes of chemical species that are broken into ions: *a) strong, water soluble acids, b) strong, water soluble bases, and c) strong, water soluble salts.*

Total Ionic Equation

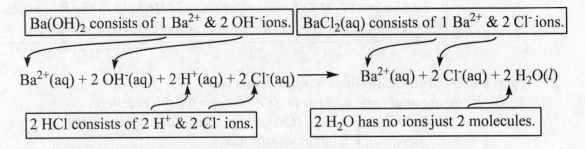

| $Ba(OH)_2$ consists of 1 Ba^{2+} & 2 OH^- ions. | $BaCl_2(aq)$ consists of 1 Ba^{2+} & 2 Cl^- ions. |

$$Ba^{2+}(aq) + 2\,OH^-(aq) + 2\,H^+(aq) + 2\,Cl^-(aq) \longrightarrow Ba^{2+}(aq) + 2\,Cl^-(aq) + 2\,H_2O(l)$$

| 2 HCl consists of 2 H^+ & 2 Cl^- ions. | 2 H_2O has no ions just 2 molecules. |

INSIGHT:

Spectator ions do not change from reactant to product. Ba^{2+} and Cl^- are spectator ions in this reaction. Once the correct total ionic equation is written, *removal of the spectator ions, Ba^{2+} and Cl^-, leaves the correct net ionic equation.*

Net Ionic Equation

$$2\,OH^-(aq) + 2\,H^+(aq) \longrightarrow 2\,H_2O(l)$$

or in the smallest possible whole numbers

$$OH^-(aq) + H^+(aq) \longrightarrow H_2O(l)$$

INSIGHT:

Just as discussed in Module 7, we reduce the stoichiometric coefficients to their smallest possible whole numbers.

6. Check your answer

a) We have used the formula unit equation to derive the total and net ionic equations.

b) We removed the spectator ions to get the net ionic equation.

c) Our answer is correct.

8. *Write the total and net ionic equations for this reaction.*

$$NaOH(aq) + CH_3COOH(aq) \rightarrow NaCH_3COO(aq) + H_2O(l)$$

The correct total ionic equation is:

$$Na^+(aq) + OH^-(aq) + CH_3COOH(aq) \rightarrow Na^+(aq) + CH_3COO^-(aq) + H_2O(l)$$

The correct net ionic equation is:

$$OH^-(aq) + CH_3COOH(aq) \rightarrow CH_3COO^-(aq) + H_2O(l)$$

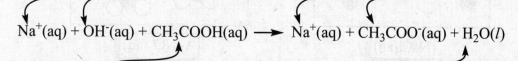

NaOH is a **strong**, water soluble base that ionizes into Na^+ and OH^- ions in aqueous solutions.	NaCH₃COO is a **strong**, water soluble salt that ionizes into Na^+ and CH_3COO^- ions in aqueous solutions.

$$Na^+(aq) + OH^-(aq) + CH_3COOH(aq) \longrightarrow Na^+(aq) + CH_3COO^-(aq) + H_2O(l)$$

CH₃COOH is a **weak**, water soluble acid that ionizes so slightly in aqueous solutions that it is not separated into ions.	H₂O is a molecule and does not form ions.

Once the Na^+ spectator ion is removed the net ionic equation remains.

$$OH^-(aq) + CH_3COOH(aq) \longrightarrow CH_3COO^-(aq) + H_2O(l)$$

Notice that the net ionic equation tells us that a strong base, hydroxide ion, reacts with a weak acid, acetic acid, to form the acetate ion and water.

9. *Draw an image that represents the following chemical reaction at the ionic/molecular level. To simplify the image, leave out the solvent water molecules.*

$$NaCl(aq) + AgNO_3(aq) \rightarrow NaNO_3(aq) + AgCl(s)$$

One possible correct image is given below:

Image Key

 = Na⁺ ●⁺ = Ag⁺ ○⁻ = Cl⁻ ● = NO₃⁻

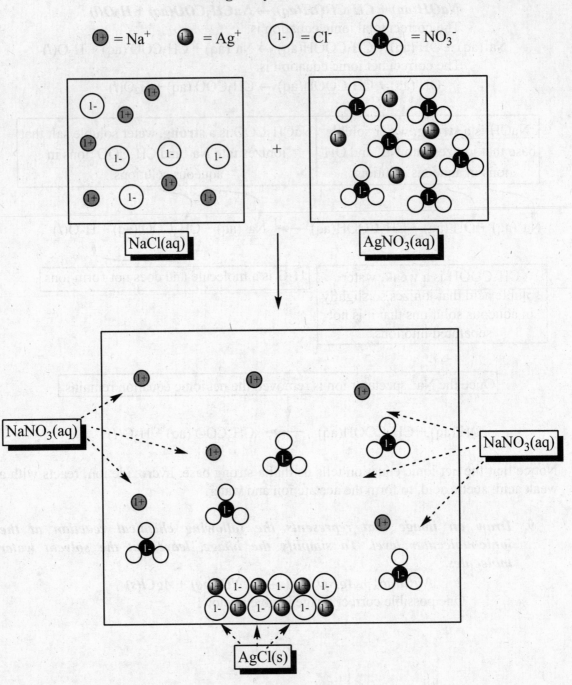

NaCl(aq)   +   AgNO₃(aq)

NaNO₃(aq)   NaNO₃(aq)

AgCl(s)

YIELD

It will improve your understanding of these reactions in solution if you draw images similar to the one in Exercise 9 for the reactions used in Exercises 4 through 8.

54

Module 9
Chemical Reaction Stoichiometry

Introduction

A balanced chemical reaction equation provides the reaction ratios for all of the species involved in the reaction. In Module 7, we saw that balanced equations are necessary to obey the Law of Conservation of Mass. In this module, we will take advantage of the Law of Conservation of Mass to determine the amounts of reactants or products involved in a chemical reaction. The important points introduced in this module are <u>a) basic b) reaction stoichiometry, c) limiting reactant calculations, and d) percent yield calculations</u>. You will need a periodic table to determine the molecular weights needed in these exercises.

Module 9 Key Equations & Concepts

1. **Limiting Reactant Concept**

    In chemical reactions, starting amounts of the reactants are finite. When one of the reactants is entirely consumed by the reaction, the reaction can not proceed, thus determining the amounts of products that can be made. In other words, the amount of one of the reactants limits the amount of products formed.

2. **Theoretical Yield**

    The amount of products calculated using the limiting reactant concept is the theoretical yield.

3. **Actual Yield**

    The amount of products experimentally formed in a reaction is the actual yield. Actual and theoretical yields frequently have different values because chemical reactions do not always produce the maximum possible amount of products.

4. **Percent Yield** $= \%$ **Yield** $= \dfrac{\textbf{Actual Yield}}{\textbf{Theoretical Yield}} \times \textbf{100\%}$

    The percent yield formula is used to determine how much of the theoretical yield was formed in a reaction.

Simple Reaction Stoichiometry Sample Exercises

1. ***How many H_2O molecules are required to exactly react with 50.0 Na atoms? How many H_2 molecules are produced by the reaction from the first step?***

    $$2\,Na \;+\; 2\,H_2O \;\rightarrow\; 2\,NaOH \;+\; H_2$$

    The correct answer is 50.0 H_2O molecules and 25.0 H_2 molecules.

    $$? \; H_2O \text{ molecules} = 50.0 \; \cancel{Na \text{ atoms}} \; \frac{2 \; H_2O \text{ molecules}}{2 \; \cancel{Na \text{ atoms}}} = 50.0 \; H_2O \text{ molecules}$$

    This conversion factor, called a reaction ratio, is derived from the reaction's stoichiometric coefficients.

    $$\underline{2}\,Na \;+\; \underline{2}\,H_2O \;\rightarrow\; 2\,NaOH \;+\; H_2$$

$$? \; H_2 \; \text{molecules} = 50.0 \; \text{Na atoms} \; \frac{1 \; H_2 \; \text{molecule}}{2 \; \text{Na atoms}} = 25.0 \; H_2 \; \text{molecules}$$

Notice that the reaction ratio is different because in the balanced reaction 1 H_2 molecule is produced for every 2 Na atoms that react.
$$2 \; Na \; + \; 2 \; H_2O \; \rightarrow \; 2 \; NaOH \; + \; 1 \; H_2$$

YIELD Reaction ratios are the key to reaction stoichiometry. Once the reaction is balanced, you can make a reaction ratio for any reactant-to-reactant, reactant-to-product, or product-to-product pair required for the problem.

2. *How many moles of H_2 can be formed from the reaction of 3.0 moles of Na with excess H_2O?*
$$2 \; Na \; + \; 2 \; H_2O \; \rightarrow \; 2 \; NaOH \; + \; H_2$$
The correct answer is 1.5 moles of H_2.

In Sample Exercise 1, we determined the numbers of atoms and molecules involved in a chemical reaction using reaction stoichiometry. In Sample Exercise 2, we are working essentially the same problem using moles. We can do this because moles are a different way of expressing the number of atoms, ions, or molecules.

INSIGHT: The word **excess** is important. It is your clue that this problem does **not** involve a limiting reactant calculation.

$$? \; \text{moles} \; H_2 = 3.0 \; \text{moles Na} \left(\frac{1 \; \text{mole of} \; H_2}{2 \; \text{moles of Na}} \right) = 1.5 \; \text{moles} \; H_2$$

Notice that the reaction ratio in this problem is exactly the same as in exercise 1 but using moles rather than atoms or molecules.

3. *How many grams of H_2 can be formed from the reaction of 11.2 grams of Na with excess H_2O?*
$$2 \; Na \; + \; 2 \; H_2O \; \rightarrow \; 2 \; NaOH \; + \; H_2$$
The correct answer is 0.494 g of H_2.

Now we make the logical step of going from moles of a substance to grams of a substance. Back in Module 6, we determined how to make the transition from moles to grams or vice versa using the molar mass of a substance. Refresh your memory if necessary.

Our solution pathway for this problem is:

grams of Na → moles of Na → reaction ratio → moles of H_2 → grams of H_2

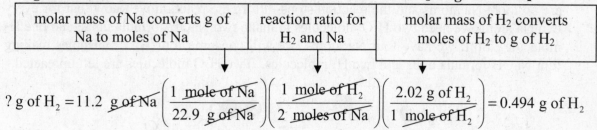

| molar mass of Na converts g of Na to moles of Na | reaction ratio for H_2 and Na | molar mass of H_2 converts moles of H_2 to g of H_2 |

$$? \text{ g of } H_2 = 11.2 \text{ g of Na} \left(\frac{1 \text{ mole of Na}}{22.9 \text{ g of Na}} \right) \left(\frac{1 \text{ mole of } H_2}{2 \text{ moles of Na}} \right) \left(\frac{2.02 \text{ g of } H_2}{1 \text{ mole of } H_2} \right) = 0.494 \text{ g of } H_2$$

This problem makes the complete transformation from grams of one of the reactants, Na, to grams of one of the products, H_2. This is a very common set of transformations that you will be expected to perform in many reaction stoichiometry problems.

Limiting Reactant Sample Exercises

4. *Using the image shown below and the chemical reaction equation, determine the maximum number of NaOH formula units and H_2 molecules that can be formed from the starting reaction mixture.*

$$2 \, Na + 2 \, H_2O \rightarrow 2 \, NaOH + H_2$$

Image Key

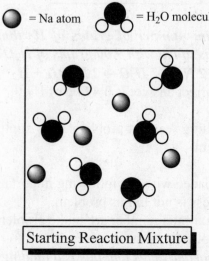

= Na atom   = H_2O molecule

Starting Reaction Mixture

The correct answer is 4 NaOH formula units and 2 H_2 molecules.

| INSIGHT: | Limiting reactants are not an unusual idea to you. In fact, you use them frequently in your everyday life. For example, a place setting for your family's Thanksgiving dinner table consists of 2 forks, 1 spoon, 1 knife, 2 plates, and 2 glasses. If you look inside the cabinets and silverware drawer to discover that there are 12 forks, 8 spoons, 7 knives, 14 plates, and 13 glasses, how many place settings can you set for the dinner table? Once you think about this for a minute, you will realize that the 12 forks determine the number of place settings at 6. As soon as that 12th fork is placed on the table, you cannot make any more place settings. It limits the number of place settings and therefore is the limiting reactant. |

In the image (see below) there are four Na atoms and six H_2O molecules. The four Na atoms are the limiting reactant. When that fourth Na atom is used to make NaOH, there are no more Na atoms and the reaction stops. From the reaction we know that for every two Na atoms we need two H_2O molecules to make two NaOH formula units, and one H_2 molecule. So if we have four Na atoms, we will consume four H_2O molecules making four NaOH formula units and two H_2 molecules. Two H_2O molecules are left unreacted.

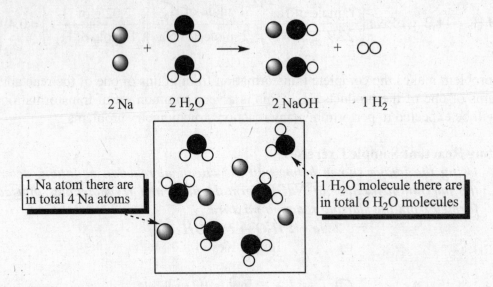

| 2 Na | 2 H_2O | | 2 NaOH | 1 H_2 |

1 Na atom there are in total 4 Na atoms

1 H_2O molecule there are in total 6 H_2O molecules

5. **What is the maximum number of grams of H_2 that can be formed from the reaction of 11.2 grams of Na with 9.00 grams of H_2O?**

$$2\ Na + 2\ H_2O \rightarrow 2\ NaOH + H_2$$

The correct answer is 0.494 g of H_2.

Because this is a typical limiting reactant problem, let's apply our problem solving steps check list to solve this problem.

1. **Read the problem**
   a) This problem is loaded with the following important information
      1) The word excess is not in this problem.
      2) The words maximum number are in the problem.
      3) Amounts of both reactants are given.
      4) These are *important clues that this is a limiting reactant problem*.
2. **Understand precisely what the problem is asking**
   a) We are being asked to determine the mass of H_2 that can be formed from fixed amounts of 2 reactants.
3. **Identify the chemical principles**
   a) This is a limiting reactant problem.
   b) We will need to use these reaction stoichiometry principles.
      1) Converting mass to moles
      2) Reaction ratios
      3) Converting moles back to mass
      4) Determining the limiting reactant
      5) Determining the maximum amount of product possible

4. **Determine the relevant pieces of information**
   a) In this problem the following information is important
      1) Maximum number – important clue that this is a limiting reactant problem
      2) Mass of Na
      3) Mass of H_2O
      4) Balanced chemical reaction equation
5. **Create a solution pathway**
   a) To determine which is the limiting reactant, perform reaction stoichiometry for each reactant having a starting amount in the problem statement.
      1) For this problem two reaction stoichiometry steps are necessary
         i. 1 step for Na
         ii. 1 step for H_2O

   b) Determine the grams of H_2 that can be made from 11.2 g of Na.

   $$? \text{ g of } H_2 = 11.2 \ \text{g of Na} \left(\frac{1 \ \text{mole of Na}}{22.9 \ \text{g of Na}} \right) \left(\frac{1 \ \text{mole of } H_2}{2 \ \text{moles of Na}} \right) \left(\frac{2.02 \ \text{g of } H_2}{1 \ \text{mole of } H_2} \right) = 0.494 \text{ g of } H_2$$

   c) Determine the grams of H_2 that can be made from 9.00 g of H_2O.

   $$? \text{ g of } H_2 = 9.00 \ \text{g of } H_2O \left(\frac{1 \ \text{mole of } H_2O}{18.0 \ \text{g of } H_2O} \right) \left(\frac{1 \ \text{mole of } H_2}{2 \ \text{moles of } H_2O} \right) \left(\frac{2.02 \ \text{g of } H_2}{1 \ \text{mole of } H_2} \right) = 0.505 \text{ g of } H_2$$

   d) Based upon the two calculations choose the maximum amount of H_2 that can be made.
      1) We can make at most 0.494 g of H_2 because it is the smaller of the two amounts we have calculated.

YIELD | **The maximum amount will be the _smallest_ amount (mass) that you calculate in the reaction stoichiometry steps!**

6. **Check your answer**
   a) We have performed a limiting reactant calculation.
   b) We have correctly calculated the possible amounts of H_2 from the two reactants.
   c) We have chosen the smallest amount H_2 possible as the correct answer.

INSIGHT: | Notice in the problem that there are 11.2 g of Na and 9.00 g of H_2O. The limiting reactant is **not** the smaller amount (9.00 g of H_2O). You can only determine the limiting reactant after you have performed the reaction stoichiometry. Trying to choose the limiting reactant based upon the amounts given in the problem is guaranteed to lead to mistakes!

Percent Yield Sample Exercise

6. *If 11.2 g of Na reacts with 9.00 g of H₂O and 0.400 g of H₂ is formed, what is the percent yield of the reaction?*

$$2\ Na + 2\ H_2O \rightarrow 2\ NaOH + H_2$$

| This is the actual yield. |

The correct answer is 81.0%.

| **INSIGHT:** | Key clues that indicate percent yield problems are a) amounts of both reactants, b) an amount for the product, and c) the words percent yield. |

In percent yield problems, limiting reactant calculations are frequently performed first to determine the theoretical yield. For this problem, the theoretical yield is the same as determined in exercise 5, 0.494 g of H₂.

$$\% \text{ yield } = \frac{\text{actual yield}}{\text{theoretical yield}} \times 100 = \frac{0.400 \text{ g}}{0.494 \text{ g}} \times 100\% = 81.0\%$$

Module 10
Electronic Structure of Atoms

Introduction

In this module we will look at early attempts to explain the line spectra of atoms, the quantum mechanical model of the atom, and some of the basic rules of quantum numbers. The important things to understand from this module are a) how line spectra occur in atoms, b) how to determine the quantum numbers for an element, c) what constitutes the valence electrons in an atom, and d) how to discern the correct atomic electronic structure of an element using the periodic table.

Module 10 Key Equations & Concepts

1. **Atomic Emission Spectrum**
   Upon excitation with energy, all atoms emit electromagnetic radiation (light of varying energy) as they release energy. Each element has a unique emission spectrum. Several elements emit that light in the visible portion of the spectrum as brightly colored lines on a black background.

2. **Atomic Absorption Spectrum**
   When elements absorb energy, their atoms generate an absorption spectrum that is the counterpart to their emission spectrum. Absorption spectra consist of black lines superimposed upon the rainbow pattern of the visible spectrum.

3. **n = 1, 2, 3, 4, 5, 6, ...∞**
   This equation defines the principal quantum number, n. The main energy level of an atom is described by its principal quantum number. For the representative elements, the principal quantum number equals the row number of the element on the periodic table.

4. **l = 0, 1, 2, 3, 4, (n-1) or l = s, p, d, f, g, ...(n-1)**
   This equation defines the orbital angular momentum quantum number, l. The shape of the atomic orbitals, the region of space occupied by electrons, is described by the orbital angular momentum quantum number.

5. **m_l - magnetic quantum number**
   The number of possible atomic orbitals for each value of l is determined by the magnetic quantum number.

6. **m_s – spin quantum number**
   The spin quantum number describes the relative magnetic orientation of the electrons in an atom. The maximum number of electrons that can occupy any one orbital is two because the spin quantum number has only two values, +1/2 and -1/2.

Emission and Absorption Spectra Sample Exercises

1. *Shown below are the emission and absorption spectra for the element hydrogen, H. Which is the emission spectrum and which is the absorption spectrum?*

   The correct answer is the top one is hydrogen's emission spectrum and the bottom one is hydrogen's absorption spectrum.

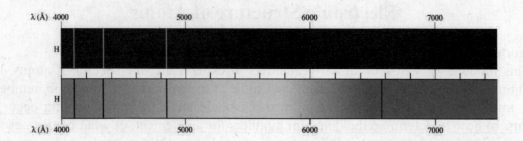

Emission spectra have black backgrounds with colored lines embedded in them. Two of hydrogen's emission lines are indicated below. Absorption spectra have rainbow colored backgrounds with black lines embedded in them. Notice that the black lines in the absorption spectrum occur at the same wavelength as the colored lines in the emission spectrum. Four emission/absorption lines are present in the H spectra shown below.

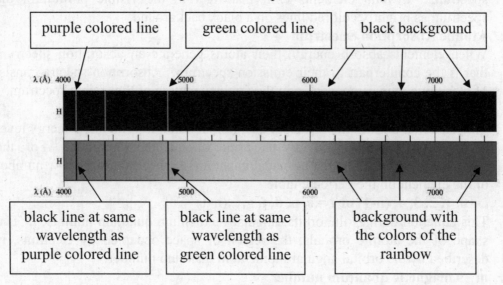

2. *How does the Bohr model of the atom explain emission and absorption spectra?*
   The explanation is emission spectra occur when electrons transition from a higher energy level in an atom to a lower energy level. Absorption spectra occur as electrons transition from lower energy levels to higher energy levels inside the atom.

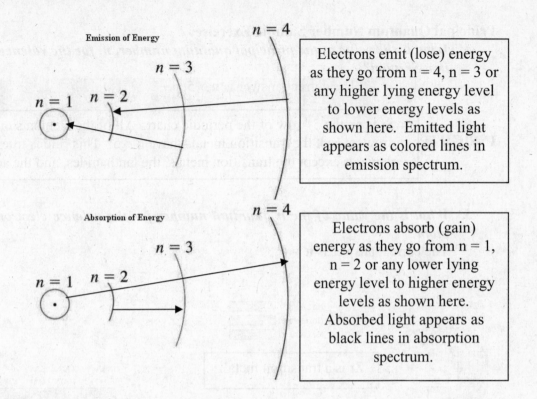

Emission of Energy

$n = 4$
$n = 3$
$n = 2$
$n = 1$

Electrons emit (lose) energy as they go from n = 4, n = 3 or any higher lying energy level to lower energy levels as shown here. Emitted light appears as colored lines in emission spectrum.

Absorption of Energy

$n = 4$
$n = 3$
$n = 2$
$n = 1$

Electrons absorb (gain) energy as they go from n = 1, n = 2 or any lower lying energy level to higher energy levels as shown here. Absorbed light appears as black lines in absorption spectrum.

Periodic Table Structure Based Upon Quantum Numbers Sample Exercises

3. *How does the shape of the periodic table exhibit the quantum numbers of each element?*

Each section of the periodic table can be associated with the **n** and *l* quantum numbers as shown below.

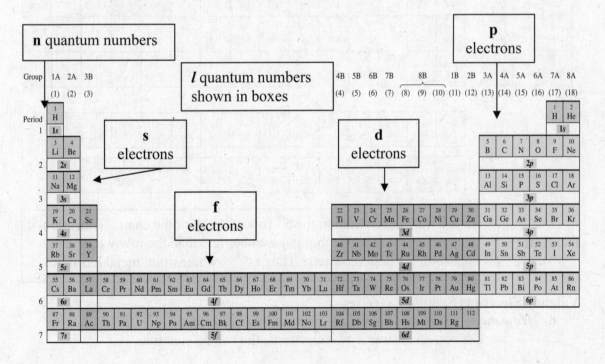

Principal Quantum Number Sample Exercises

4. *What is the value of the principal quantum number, n, for the valence electrons in a Sr atom?*

   The correct answer is **n** = 5.

INSIGHT:	Sr is on the 5th row of the periodic chart. All of the elements on the 5th row, except the transition metals, have n = 5. This rule is true for all elements except the transition metals, the lanthanides, and the actinides.

5. *What is the value of the n quantum number for the valence electrons in a Zr atom?*

   The correct answer is **n** = 4.

Alkali metals

Metals

Nonmetal

Metalloi

Noble gases

1A (1)

Alkaline earth metals

8A (18)

Halogens

Zr is a transition metal.

Transition metals

INSIGHT:	Zr is a transition metal on the 5th row of the periodic chart. Transition metals have an **n** value that is 1 number less than the row where they appear on the periodic chart. Thus a 5th row transition metal has **n** = 4.

Valence Electrons Sample Exercises

6. *How many valence electrons are present in a Sr atom?*

   The correct answer is two electrons, the two 5s electrons.

7. **How many valence electrons are present in a Zr atom?**

   The correct answer is four electrons, the two 5**s** electrons and the two 4**d** electrons.

> **INSIGHT:** In transition metals, both **s** and **d** electrons can form chemical bonds. Thus for Zr there are four valence electrons, two 5s electrons and two 4d electrons.

Orbital Angular Momentum Quantum Number Sample Exercises

8. **What is the value of the orbital angular momentum quantum number, l, for the valence electrons in a Sr atom?**

   The correct answer is $l = 0$ or s electrons.

   For Sr which has $n = 5$, l can be 0, 1, 2, 3, 4 or s, p, d, f, g electrons.

> **INSIGHT:** From the key equation for the l quantum number, we know that $l = 0, 1, 2, 3, \ldots(n\text{-}1)$. Sr has $n = 5$, so $n\text{-}1 = 4$. Thus we know that for Sr $l = 0, 1, 2, 3, 4$ which is equivalent to saying that in a Sr atom we can find **s** electrons ($l = 0$), **p** electrons ($l = 1$), **d** electrons ($l = 2$), **f** electrons ($l = 3$), and **g** electrons ($l = 4$)

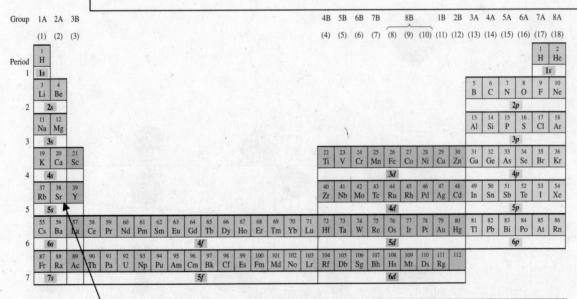

> **INSIGHT:** Notice that the two electrons which make Sr, element 38, different from Ar, element 36, are in the **s** block of the periodic table. *All s electrons have an l value of 0.*

9. **What is the value of the orbital angular momentum quantum number, *l*, for the electrons which make a Zr atom different from a Sr atom?**

The correct answer is $l = 2$ or d electrons.

For Zr which has $n = 4$, l can be 0, 1, 2, 3 or **s, p, d, f** electrons.

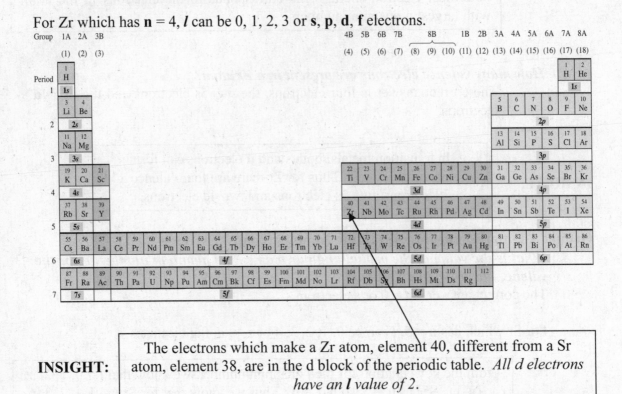

INSIGHT: The electrons which make a Zr atom, element 40, different from a Sr atom, element 38, are in the d block of the periodic table. *All d electrons have an l value of 2.*

10. **Choose the picture of the atomic orbitals which have an *l* value of 1.**

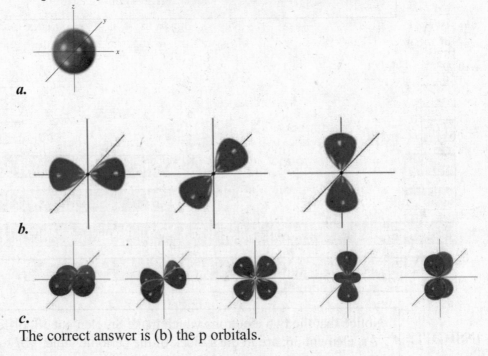

a.

b.

c.

The correct answer is (b) the p orbitals.

Answer (a) is a representation of an **s** orbital. All **s** orbitals have $l = 0$. Answer (b) is a representation of the three **p** orbitals found in a given shell. All **p** orbitals have $l = 1$. Answer (c) is a representation of the five **d** orbitals found in a given shell. All **d** orbitals have $l = 2$.

Magnetic Quantum Number Sample Exercise

    11. How many d orbitals are present in an atomic energy level that contains d orbitals?

                The correct answer is five d orbitals.

There are always at most five d orbitals in any one atomic energy level. A picture of the five **d** orbitals is shown in exercise 10 as answer c.

INSIGHT:	In any one energy level there are at most:
	  a) one **s** orbital
	  b) three **p** orbitals
	  c) five **d** orbitals

Spin Quantum Number Sample Exercise

    12. What is the maximum number of electrons that can exist in d orbitals in one atomic energy level?

    The correct answer is 10.

Because there are five **d** orbitals and each **d** orbital can contain at most two electrons (one spin up and the other spin down) a total of 10 electrons are the most that can occupy **d** orbitals.

INSIGHT:	Each orbital can hold at most two electrons. Thus in:
	  a) **s** orbitals have 1 orbital $\times$ 2 electrons = 2 electrons total
	  b) **p** orbitals have 3 orbitals $\times$ 2 electrons = 6 electrons total
	  c) **d** orbitals have 5 orbitals $\times$ 2 electrons = 10 electrons total

Electronic Structure from the Periodic Chart Sample Exercises

    13. What is the correct electronic structure of the Sr atom? Write the structure in both orbital notation and simplified (or spdf) notation.

        The correct answer is [Kr] $\uparrow\downarrow$ or [Kr] $5s^2$.
                                  5s

This is the appropriate exercise to use the problem solving steps check list to help us determine the correct answer.

    1. **Read the problem**
        a) Notice that the problem wants the electronic structure of a Sr atom written in two different formats.
    2. **Understand precisely what the problem is asking**

b) We are being asked to determine the electronic structure of Sr using both orbital and spdf notations.

3. Identify the chemical principles

c) Understanding how to assign electronic structures for elements is the chemical principle involved.

4. Determine the relevant pieces of information

a) The most important pieces of information are that the electronic structure of Sr is required.

b) We must write this in both orbital and spdf notations.

5. Create a solution pathway

YIELD

The easiest way to write electron configurations for elements is to use the modified periodic table shown below. You must find the element of interest then comprehend that element's row and electron block.

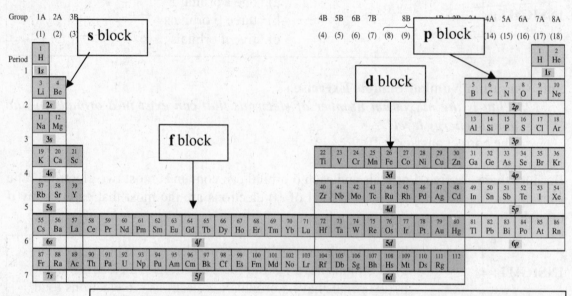

INSIGHT: The element's noble gas core configuration is determined by starting at the element then decreasing the atomic number (counting backward on the periodic table) until reaching a noble gas. That noble gas's symbol is used to represent the element's non-valence (core) electrons

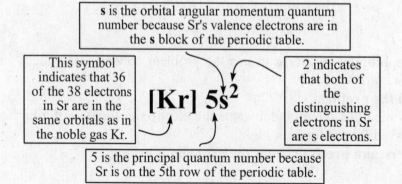

s is the orbital angular momentum quantum number because Sr's valence electrons are in the s block of the periodic table.

This symbol indicates that 36 of the 38 electrons in Sr are in the same orbitals as in the noble gas Kr.

2 indicates that both of the distinguishing electrons in Sr are s electrons.

$$[Kr]\ 5s^2$$

5 is the principal quantum number because Sr is on the 5th row of the periodic table.

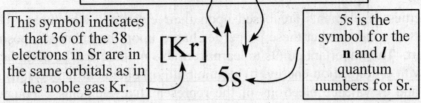

The ↑ indicates that the spin quantum number for one of the valence electrons is +1/2. The ↓ symbolizes that the spin quantum number is -1/2 for the second valence electron.

This symbol indicates that 36 of the 38 electrons in Sr are in the same orbitals as in the noble gas Kr.

$[Kr]$ ↑↓
5s

5s is the symbol for the **n** and *l* quantum numbers for Sr.

6. Check your answer

a) We have used the modified periodic table to determine the electronic structure of Sr.

b) We have written the electronic structure in both orbital and spdf notation.

c) Our answer is correct.

14. What is the correct electronic structure of the Zr atom? Write the structure in simplified (or spdf) notation.

The correct answer is $[Kr] \, 5s^2 \, 4d^2$.

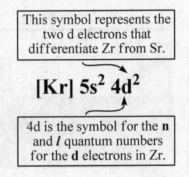

This symbol represents the two d electrons that differentiate Zr from Sr.

$$[Kr] \, 5s^2 \, 4d^2$$

4d is the symbol for the **n** and *l* quantum numbers for the **d** electrons in Zr.

Module 11
Periodic Trends in Chemistry

Introduction

Many properties of elements are based upon their electronic structures. Using some simple trends, we can predict these properties based upon the element's position on the periodic chart. The important points to learn from this module are the periodic properties associated with a) ionization energy, b) atomic radii, c) chemical families, d) metals and nonmetals, and e) valence electrons of the representative elements. You will need a periodic chart with you as you work on this module to learn to associate these properties with the periodic chart.

Module 11 Key Equations & Concepts

1. **Ionization Energy**

    Every element can have one or more electrons removed from their atoms. The amount of energy required to remove the first electron from an atom is called the first ionization energy. Ionization energy is an important indicator of an element's likelihood of forming positive ions. The trend for ionization energy is indicated below.

    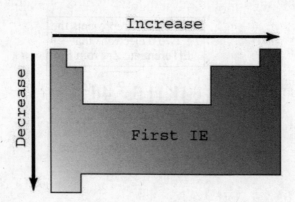

2. **Atomic Radii**

    Every atom of an element has a measurable volume. Since atoms approximate spheres in shape, if we know an element's atomic radius we also know its volume. Atomic radii are the measured distance from the center of the atom to its outer electrons. The trend for atomic radii is indicated below.

    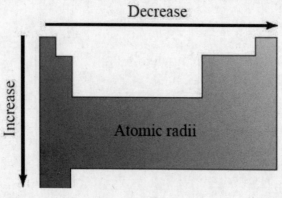

---

3. Chemical Families
Representative elements in a column of the periodic table have many similar properties because of their electronic structures. Representative elements in a column on the periodic table are called a family.

4. Valence Electrons
Those electrons involved in chemical bond formation are called valence electrons. For representative elements, the valence electrons are those with the highest n quantum number value for their given row of the periodic table.

Ionization Energy Sample Exercise

1. *Arrange these elements from smallest to largest based on their ionization energies.*

F, N, C, O

The correct answer is C < O < N < F.

INSIGHT: First ionization energies steadily increase from the alkali metals to the noble gases. There are two significant variations from this general trend. Elements in the IIA and VA columns have larger ionization energies than expected from the trend because IIA elements have filled **s** orbitals and VA elements have half-filled **p** orbitals.

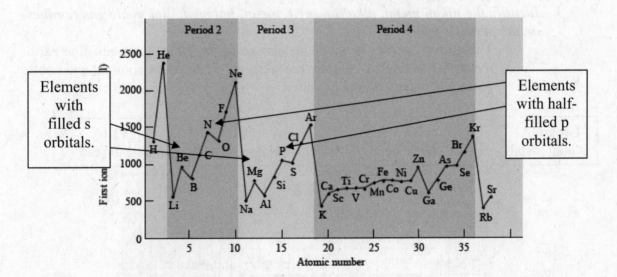

Atomic Radii Sample Exercise

2. *Arrange these elements from smallest to largest based on their atomic radii.*

F, Ga, S, Rb

The correct answer is F < S < Ga < Rb.

INSIGHT: There are two important trends to remember about atomic radii. 1) Atomic radii increase from the left-side to the right-side of the periodic table. 2) Atomic radii increase from the top of the periodic table to the bottom.

71

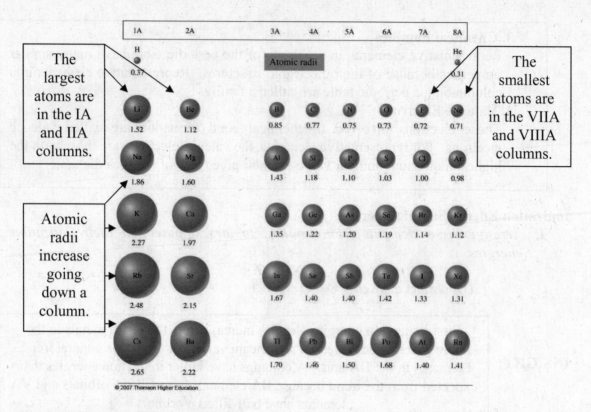

Chemical Families Sample Exercise

3. *Identify the alkali metal, alkaline earth metal, halogen, and noble gas families on the periodic table.*

The correct answer is alkali metals are in the IA column, alkaline earth metals are in the IIA column, halogens are in the VIIA column, and noble gases are in the VIIIA column.

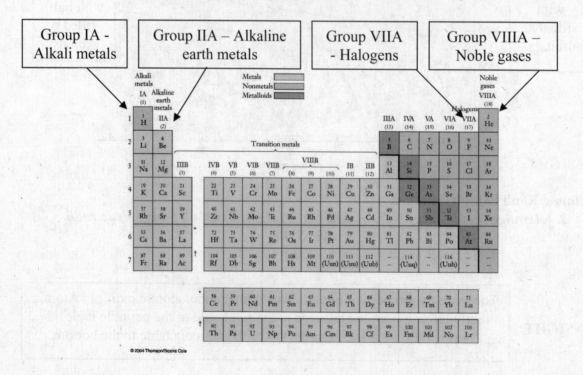

Frequently, your instructor will refer to these four chemical families. You will need to know which elements are in each family and to recognize that each member of the family has common reactivity characteristics. For example, all of the alkali metals rapidly react with air and water. All of the noble gases are unreactive to nearly every other element except under the harshest of conditions.

Metals and Nonmetals Sample Exercise

4. *Identify the positions of the metals, nonmetals, and metalloids on the periodic table.*

The correct answer is that metals occupy the left 80% of the periodic table. Nonmetals occupy the right portion of the periodic table. Metalloids are the elements that divide the metals from the nonmetals.

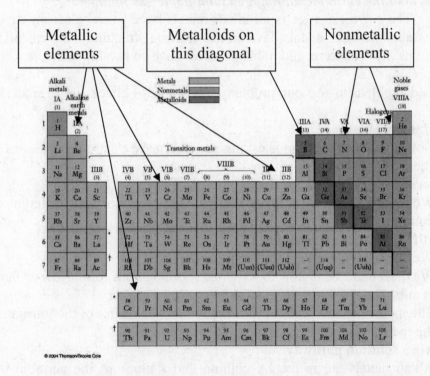

From a chemical reactivity perspective, metals tend to easily lose electrons forming positive ions when they react with nonmetals. Nonmetals tend to gain electrons forming negative ions in their reactions with nonmetals. Metalloids have an intermediate reactivity between metals and nonmetals.

5. *Which elements have the largest metallic character?*

The correct answer is the most metallic elements are rubidium, cesium, strontium, and barium.

Metallic character is largest for elements in the lower left portion of the periodic table and smallest for elements in the top right portion of the periodic table as indicated in the graph below. Fr and Ra are highly radioactive metallic elements that are difficult to use in typical labs.

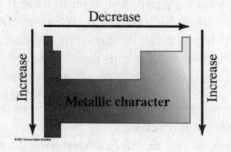

Decrease →

↓ Increase

↑ Increase

Metallic character

© 2007 Thomson Higher Education

Valence Electrons Sample Exercises

6. *Write the valence electron structures in condensed (spdf) notation for the alkali metal, alkaline earth metal, halogen, and noble gas families.*

> The correct answer is all alkali metals have an ns^1 electron structure, all alkaline earth metals have an ns^2 electron structure, halogens all have the $ns^2 \, np^5$ structure, and noble gases all have an $ns^2 \, np^6$ structure.

This is a good problem to use our problem solving steps check list to understand the answers.

1. **Read the problem**
    a) Notice that this question is asking us for generic electronic notations for these chemical families.
2. **Understand precisely what the problem is asking**
    a) We are being asked to write the spdf notations for four different chemical families.
3. **Identify the chemical principles**
    a) We must recognize the four chemical families.
    b) We must understand how to write the spdf notations for these four families.
4. **Determine the relevant pieces of information**
    a) The most relevant pieces of information are the names of the four families and the indication that we must write their spdf notations.
5. **Create a solution pathway**
    a) Alkali metals are in the IA column and **s** block of the periodic table (see Module 10). The IA designation for the family column indicates that there is one valence electron for each of these elements. Thus each one of the alkali metals has an s^1 structure. For example, Na is in the 3rd row of the periodic table and its correct electron structure is $[Ne]3s^1$. By comparison, Cs is in the 6th row giving a $[Xe]6s^1$ structure.
    b) Alkaline earth metals have ns^2 electronic structures. All alkaline earth metals are in the IIA column (have two valence electrons) and **s** block of the periodic table. For example, Ca (in the 4th row of the periodic table) has an $[Ar]4s^2$ electronic structure.
    c) Halogens occupy the VIIA column and thus have seven valence electrons. Halogens are also in the p block of the periodic table. The correct way to write that structure for the 5th row element I is $[Kr]5s^2 \, 5p^5$.

74

d) Noble gases (except for He) all have eight valence electrons, two in the **s** orbital and six in the **p** orbitals. (He only has two electrons in the **s** orbital.) Ne in the 2nd row of the periodic table has a [He]2$\mathbf{s}^2$ 2$\mathbf{p}^6$ structure.

6. **Check your answer**
   a) Alkali metals, IA column and **s** block, have n$\mathbf{s}^1$ structure.
   b) Alkaline earth metals, IIA column and **s** block, have n$\mathbf{s}^2$ structure.
   c) Halogens, VIIA column and **p** block, have n$\mathbf{s}^2$ n$\mathbf{p}^5$ structure.
   d) Noble gases, VIIIA column and **p** block, have n$\mathbf{s}^2$ n$\mathbf{p}^6$ structure.
   e) Our answers are correct.

7. *How many valence electrons are present in an As atom? Write the electronic structure of an As atom.*

   The correct answer is there are five valence electrons in an As atom. The electronic structure for these five electrons is [Ar]4$\mathbf{s}^2$ 4$\mathbf{p}^3$.

Arsenic is in the VA column of the periodic table. This indicates that As has five valence electrons. Arsenic is in the **p** block and on the 4th row of the periodic table. Thus we know that the correct electronic structure is [Ar]4$\mathbf{s}^2$ 4$\mathbf{p}^3$.

YIELD

Chemical bonding is based upon the electronic structures of the elements. To understand how elements bond to form chemical compounds, you must understand how many valence electrons in which orbitals are present for any representative element. To easily determine these values, use the family column numbers for the number of valence electrons and the blocks of the periodic table for the electronic structure.

Module 12
Chemical Bonding

Introduction

This module is concerned with how atoms combine to form chemical bonds. There are two basic types of chemical bonds, ionic and covalent. This module's important points are <u>a) how to determine if a compound is ionic or covalent, b) drawing Lewis dot formulas of atoms, c) writing formulas of simple ionic compounds, d) drawing Lewis dot structures of ionic and covalent compounds, e) electronegativity, and f) recognizing if a covalent bond is polar or nonpolar.</u> We will frequently use the electron configurations of the elements as previously described in Module 10. Refer there if necessary as you work on this module.

Module 12 Key Equations & Concepts

1. **Lewis Dot Formulas**

   Chemists frequently use Lewis dot formulas as a bookkeeping method for valence electrons. Lewis dot formulas indicate both the number of valence electrons and how they are involved in chemical bonds.

2. **Ionic Bonds**

   Simple ionic bonds are formed when metals react with nonmetals. In ionic bonds, electrons are transferred from an atom of one element to the atom of another element forming positively and negatively charged ions. The actual ionic bond is the attractive force between oppositely charged ions.

3. **Covalent Bonds**

   Simple covalent bonds are formed when nonmetals react with nonmetals. In covalent bonds, electrons are shared in the valence shells of each atom. Covalent bonds center electrons between the two nuclei permitting them to approach one another and stabilize their interactions.

4. **Octet Rule**

   In the formation of many simple chemical compounds, both ionic and covalent, atoms attain an electron configuration that is equivalent to that of the nearest noble gas.

5. **Electronegativity**

   Elements do not share their electrons equally in covalent bonds. Electronegativity is a relative measure of the tendency for an atom to lose or gain electrons in a chemical bond.

6. **Nonpolar and Polar Bonds**

   Covalent bonds in which the electrons are equally shared between the two atoms are called nonpolar. Nonpolar bonds occur between two atoms that have nearly the same electronegativity. Polar covalent bonds occur when the electrons are not equally shared between the two atoms involved in bond formation. The atom with the larger electronegativity has greater control over the shared electrons and acquires a partial negative charge.

Determining if a Compound is Ionic or Covalent Sample Exercise

1. **Indicate which of the following compounds are ionic in nature and which are covalent in nature.**

$$CH_4, KBr, Ca_3N_2, Cl_2O_7, H_2SO_4, NH_4Cl$$

The correct answer is KBr, Ca_3N_2 and NH_4Cl are ionic while CH_4, Cl_2O_7, and H_2SO_4 are covalent.

INSIGHT: Ionic compounds are formed by the reaction of metallic elements with nonmetallic elements or the ammonium ion, NH_4^+, with nonmetals.

K is a metallic element.	Ca is a metallic element.	NH_4^+ is an ion.

KBr **Ca₃N₂** **NH₄Cl**

Br is a nonmetallic element.	N is a nonmetallic element.	Cl is a nonmetallic element.

INSIGHT: Covalent compounds are formed by the reaction of nonmetals with nonmetals.

C is a nonmetal.	Cl is a nonmetal.	H and S are nonmetals.

CH₄ **Cl₂O₇** **H₂SO₄**

H is a nonmetal.	O is a nonmetal.	O is a nonmetal.

Lewis Dot Formulas of Atoms Sample Exercises

2. **Draw the correct Lewis dot formula of each of these elements.**

Mg, P, S, Ar

The correct answer is:

INSIGHT: Lewis dot formulas reflect the atomic electronic structures of the valence electrons for each element, including whether the valence electrons are paired or unpaired, one dot per electron. In the following diagram, notice how the Lewis dot formulas reflect the electronic structure orbital diagrams for each element.

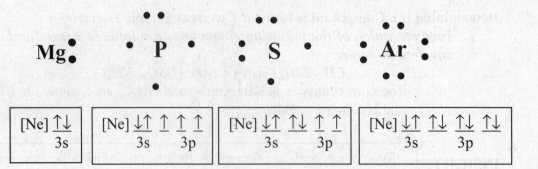

| [Ne] ↑↓
 3s | [Ne] ↑↓ ↑ ↑ ↑
 3s   3p | [Ne] ↑↓ ↑↓ ↑ ↑
 3s   3p | [Ne] ↑↓ ↑↓ ↑↓ ↑↓
 3s   3p |

> **YIELD**
>
> In the remainder of this module, we will frequently use Lewis dot formulas of various elements to describe their chemical bonding. It is important that you understand how Lewis dot formulas are related to an element's electron structure. Be certain that you know how to perform this vital skill.

Simple Ionic Compounds Sample Exercise

3. *Write the correct formulas of the ionic compounds formed when a) Mg atoms react with Cl atoms, b) Mg atoms react with S atoms, c) Mg atoms react with P atoms.*

The correct answers are: $MgCl_2$, MgS, and Mg_3P_2.

> Mg, like all of the IIA metals, has two electrons in its valence shell and commonly forms +2 ions, Mg^{2+}.

$MgCl_2$	MgS	Mg_3P_2
Cl, and all of the VIIA nonmetals have seven electrons in their valence shell commonly forming 1- ions, such as Cl^-. Two Cl^- ions have the same size charge as one Mg^{2+} ion. To form a neutral compound, there must be one Mg^{2+} ion and two Cl^- ions giving the formula $MgCl_2$.	S, and all of the VIA nonmetals have six electrons in their valence shell commonly forming 2- ions, such as S^{2-}. One S^{2-} ion has the same size charge as one Mg^{2+} ion. To form a neutral compound there must be one Mg^{2+} ion and one S^{2-} ion giving the formula MgS.	P, and all of the VA nonmetals have five electrons in their valence shell commonly forming 3- ions, such as P^{3-}. Two P^{3-} ions have the same size charge as three Mg^{2+} ions. To form a neutral compound there must be three Mg^{2+} ion and two P^{3-} ions giving the formula Mg_3P_2.

Writing correct formulas of ionic compounds is an essential skill for your success in this course. Here are two steps for you to follow to write correct ionic formulas. 1) Determine the ionic charge of the elements. IA elements form 1+ ions, IIA elements form 2+ ions, and IIIA elements form 3+ ions. VA nonmetals form 3- ions, VIA nonmetals form 2- ions, and VIIA nonmetals form 1- ions. 2) Adjust the number of positive and negative ions to ensure that the total size of the positive charge equals the total size of the negative charge as shown in exercise 3.

Drawing Lewis Dot Structures of Ionic Compounds Sample Exercise

4. *Draw the Lewis dot structures for each of these compounds.*
   AlP, NaCl, MgCl₂

The correct answers are:

$$Al^{3+} \left[: \ddot{P} :\right]^{3-} \qquad Na^+ \left[: \ddot{Cl} :\right]^- \qquad Mg^{2+} 2\left[: \ddot{Cl} :\right]^-$$

Al, a IIIA element, forms a 3+ ion by losing all of its valence electrons leaving no dots.

$$Al^{3+} \left[: \ddot{P} :\right]^{3-}$$

P, a VA element, gains three electrons from Al forming a 3- ion, and thus has 8 dots (5 valence electrons plus 3 from Al). The [ ]'s indicate that the 3- charge is associated with the P.

Na, a IA element, forms a 1+ ion losing its one valence electron leaving no dots.

$$Na^+ \left[: \ddot{Cl} :\right]^-$$

Cl, a VIIA element, gains one electron from Na forming a 1- ion and has 8 dots (7 valence electrons plus 1 from Na).

Mg, a IIA element, loses both valence electrons forming a 2+ ion, with no dots.

$$Mg^{2+} 2\left[: \ddot{Cl} :\right]^-$$

Each Cl atom gains one electron from the Mg. The 2 in front of the [ ]'s indicates that two Cl⁻ ions are needed to have the same size charge as one Mg²⁺.

Simple Covalent Compounds Sample Exercises

  5. *Draw the correct Lewis dot structures for each of these compounds.*

  SiH₄, PCl₃, SF₆

The correct answers are:

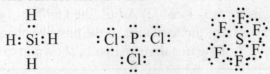

H H:Si:H H	:Cl:P:Cl: 　:Cl:	F S F arrangement

INSIGHT:	Many, but not all, chemical compounds obey the octet rule. Each element in the compound attains an electron configuration similar to the nearest noble gas. For most elements that is 8 electrons. For H it is 2.

Si and H both obey the octet rule. Si has a share of 8 electrons in 4 *bonding pairs*. Each H has a share of 2 electrons also *in bonding pairs*. This compound has only *bonding pairs* of electrons.

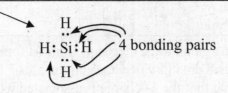

H: Si:H ⟶ 4 bonding pairs

In PCl₃, P has a share of 8 electrons arranged as 3 *bonding pairs* and 1 *lone pair* of electrons. Each Cl also has a share of 8 electrons arranged as 1 *bonding pair* and 3 *lone pairs*.

lone pair   :Cl:P:Cl: ⟶ 3 bonding pairs

INSIGHT:	When drawing Lewis dot structures, if the compound obeys the octet rule, the central atom will have a share of 8 electrons. The possible combinations of 8 electrons for compounds that **obey the octet rule** are:

Bonding Pairs	Lone Pairs
4	0
3	1
2	2
1	3

SF$_6$ does not obey the octet rule.
The S atom has a share of 12 electrons while each F has a share of 8 electrons. This compound has 6 *bonding pairs* of electrons. Look in your textbook for the rules on which compounds do not obey the octet rule.

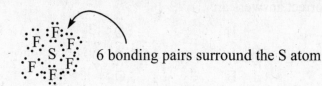

6 bonding pairs surround the S atom

If the compound **does not obey the octet rule**, the central atom can have 2, 3, 5 or 6 pairs of electrons around the central atom. All of the possible combinations of 2, 3, 5, or 6 pairs of electrons are:

Total Pairs of Electrons	Bonding Pairs	Lone Pairs
2	2	0
3	3	0
5	5	0
5	4	1
5	3	2
5	2	3
6	6	0
6	5	1
6	4	2

INSIGHT: Even if the central atom does not obey the octet rule, noncentral atoms will obey the octet rule having either 1 bonding pair, as for H atoms, or 8 electrons as is the case for F in the SF$_6$ example above.

YIELD One method to determine the number of bonding and lone pairs of electrons in a compound that obeys the octet rule is the **N-A = S** rule. N is the number of electrons NEEDED. Determine N by adding 8's for each element except H which uses 2. For PCl$_3$, N = 8 (for P) + 3x8 (for Cl) = 32. A is the number of electrons AVAILABLE. Determine A by adding the group numbers for each element in the compound. For PCl$_3$, P is in the VA group and Cl is in the VIIA group. N = 5 + 3(7) = 26. S is the number of electrons SHARED. For PCl$_3$, S = 32-26 = 6. Notice that in PCl$_3$ there are 3 shared pairs for 6 shared electrons. Lone electrons are determined from A-S which for PCl$_3$ is 26-6 = 20 electrons in lone pairs.

INSIGHT: In the N-A=S rule, for positive ions <u>decrease</u> the number of electrons available by the ionic charge. For negative ions, <u>increase</u> the number of electrons available by the ionic charge.

81

6. **Draw the correct Lewis dot structures for each of these ions.**

$$NH_4^+ , PO_4^{3-}$$

The correct answers are:

$$
\begin{array}{cc}
\text{H} \quad + & \qquad :\!\overset{\displaystyle ..}{\text{O}}\!: \quad 3\text{-} \\[4pt]
\text{H}\!:\!\overset{..}{\underset{..}{\text{N}}}\!:\!\text{H} & :\!\overset{..}{\text{O}}\!:\!\overset{..}{\underset{..}{\text{P}}}\!:\!\overset{..}{\text{O}}\!: \\[4pt]
\text{H} & \qquad :\!\overset{..}{\underset{..}{\text{O}}}\!:
\end{array}
$$

INSIGHT: For the NH_4^+ ion, N = 8 (for N) + 4x2 (for H) = 16, A = 5 (for N) + 4x1 (for H) – 1 (for the ionic charge) = 8, and S = 16 – 8 = 8 electrons in bonding pairs. For the NH_4^+ ion, electrons in lone pairs = A - S = 8 - 8 = 0. Based upon these numbers, the following Lewis dot structure is correct.

$$
\begin{array}{c}
\text{H} \quad + \\[4pt]
\text{H}\!:\!\overset{..}{\underset{..}{\text{N}}}\!:\!\text{H} \\[4pt]
\text{H}
\end{array}
$$

0 lone pairs

INSIGHT: For the PO_4^{3-} ion, N = 8 (for P) + 4x8 (for O) = 40, A = 5 (for P) + 4x6 (for O) + 3 (for the ionic charge) = 32, and S = 40 – 32 = 8 bonding electrons. For the PO_4^{3-} ion, electrons in lone pairs = A - S = 32 - 8 = 24. Count the bonding and lone electrons in this Lewis dot structure to see if it is correct.

$$
\begin{array}{c}
:\!\overset{..}{\text{O}}\!: \quad 3\text{-} \\[4pt]
:\!\overset{..}{\text{O}}\!:\!\overset{..}{\underset{..}{\text{P}}}\!:\!\overset{..}{\text{O}}\!: \\[4pt]
:\!\overset{..}{\underset{..}{\text{O}}}\!:
\end{array}
$$

7. **Draw the correct Lewis dot structures for each of these compounds.**

$$C_2H_4, CN^-$$

The correct answers are:

$$
\begin{array}{cc}
\text{H}\overset{..}{} \quad \overset{..}{}\text{H} & \\
\overset{..}{\text{C}}::\overset{..}{\text{C}} & \qquad :\text{C}:::\text{N}:^- \\
\text{H} \quad \text{H} &
\end{array}
$$

INSIGHT: For C_2H_4, N = 2x8 (for C) + 4x2 (for H) = 24, A = 2x4 (for C) + 4x1 (for H) = 12, and S = 24 – 12 = 12 bonding electrons. Electrons in lone pairs = A - S = 12 - 12 = 0. To have 12 bonding electrons with this few atoms, we must have a double bond between the two C atoms as shown below.

4 electrons between 2 atoms
indicates a double bond

Let's use our problem solving steps check list to draw the Lewis dot structure of the CN⁻ ion.

1. **Read the problem**
   a) We are to draw the Lewis dot structure of the CN⁻ ion.
2. **Understand precisely what the problem is asking**
   a) This problem is asking us to draw the Lewis dot structure of a negative ion that probably contains multiple bonds.
   b) We can tell multiple bonds are involved because C and N have 4 and 5 valence electrons, respectively, which must be shared between only two atoms.
3. **Identify the chemical principles**
   a) Drawing Lewis dot structures is the main principle but we must also know:
      1) How to determine the number of valence electrons for elements.
      2) To add the extra electron available from the ionic charge.
      3) How to determine the number of shared and lone electrons.
4. **Determine the relevant pieces of information**
   a) The chemical formula of the CN⁻ ion is the most important piece of information.
5. **Create a solution pathway**
   a) Determine the number of electrons NEEDED
      1) N = 1x8 (for C) + 1x8 (for N) = 16
   b) Determine the number of electrons AVAILABLE
      1) A = 1x4 (for C) + 1x5 (for N) + 1 (for ionic charge) = 10
   c) Determine the number of electrons SHARED
      1) N - A = S
      2) 16 – 10 = 6
   d) Determine the number of lone electrons
      1) A – S = lone electrons
      2) 10 - 6 = 4
   e) Because there are 6 shared electrons and only 2 atoms, there must be a triple bond in this molecule.
      1) Place 6 paired dots between the C and N atoms
         C ⋮⋮⋮ N
   f) We still must place the 4 lone electrons on our ion
      1) The C atom has a share of 6 electrons in our present structure.
      2) C needs 8 electrons to have an electronic structure like that of Ne.
      3) Add 2 electrons to the C atom.
         : C ⋮⋮⋮ N

g) The C atom has a completed octet but the N atom still only has a share of 6 electrons and we have 2 electrons left from the 4 lone electrons to place on the structure
   1) Add the last 2 lone electrons to the N atom and include the ionic charge

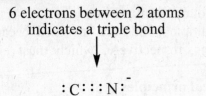

6. **Check your answer**
   a) The structure has a total of 10 electrons.
   b) There are 6 shared electrons.
   c) There are 4 lone electrons.
   d) C has a share of 8 electrons.
   e) N has a share of 8 electrons.
   f) The structure is correct.

6 electrons between 2 atoms
indicates a triple bond

$\downarrow$

$:C:::N:^{-}$

Electronegativity Sample Exercise
8. *Arrange these elements from the least electronegative to the most electronegative.*

   F, Mo, S, K

   The correct answer is K < Mo < S < F.

INSIGHT: Electronegativity indicates which element has greater control of the shared electrons in a covalent bond. The larger the electronegativity value, the greater share that element has of the electrons. The periodic trend for electronegativity is shown below.

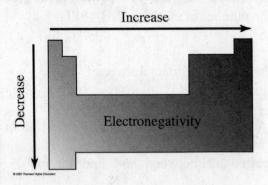

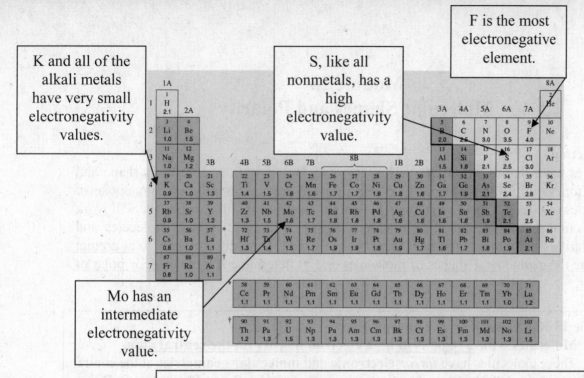

K and all of the alkali metals have very small electronegativity values.

S, like all nonmetals, has a high electronegativity value.

F is the most electronegative element.

Mo has an intermediate electronegativity value.

YIELD

Electronegativity is an important indicator of how electrons are shared in covalent bonds. If the two elements involved in a covalent bond have similar electronegativity values, the electrons are equally shared and the bond is considered to be nonpolar. If the two elements involved in a covalent bond have very different electronegativity values, the more electronegative element has the greater share of the electrons. In such cases the covalent bond is designated as a polar covalent bond.

Polar or Nonpolar Covalent Bonds in Compounds Sample Exercise

9. *Which of these compounds contains polar covalent bonds?*
   $$F_2, CH_4, H_2O$$
   The correct answer is CH_4 and H_2O contain polar covalent bonds and F_2 does not.

INSIGHT:

Polar bonds occur when the two atoms involved in the bond have a difference in electronegativity. In F_2, the two atoms are both F. They have the same electronegativity thus there is not a polar bond. In CH_4 and H_2O, the H to the central atom (C or O) bond involves atoms with different electronegativities. Consequently, there are polar covalent bonds in both CH_4 and H_2O.

This module introduces two very important concepts, Lewis dot formulas and bond polarity, which are used frequently in subsequent parts of the course. The next major subject of this course is to introduce molecular shapes and molecular polarity. To understand shapes and polarity requires merging the ideas of Lewis dot formulas and bond polarity with a theory of molecular shape which is introduced in the next module. Make sure that you understand Module 12 to ensure your success in the next part of the course.

Module 13
Molecular Shapes and Polarity

Introduction
Molecules have shapes that affect their reactivity. Based upon a molecule's shape and the electronegativities of the bonded elements, we can predict if a molecule is nonpolar or polar (does or does not have a symmetric electron density). In this module, we will work on predicting and understanding the stereochemistry of several simple molecules and their polarities. The most important ideas to take from this module are how to <u>a) predict and name the geometrical shapes of molecules and b) determine if a molecule is polar or nonpolar</u>.

Module 13 Key Equations & Concepts

1. **Molecules with <u>2 regions of high electron density on the central atom</u>**
   These molecules have *linear* electronic and **molecular geometries** at the central atom with $180°$ bond angles. Examples include BeF_2, BeH_2, $BeCl_2$, and $BeBr_2$.

2. **Molecules with <u>3 regions of high electron density on the central atom</u>**
   These molecules have *trigonal planar* electronic and **molecular geometries** at the central atom with $120°$ bond angles. Examples include BH_3, BF_3, $AlCl_3$, and GaI_3.

3. **Molecules with <u>4 regions of high electron density on the central atom</u>**
   These molecules have *tetrahedral* electronic geometry and **can have either tetrahedral, trigonal pyramidal, angular, or linear molecular geometries** at the central atom. Bond angles vary with the molecular geometries. Examples include CH_4, SiH_4, PF_3, H_2O, and HBr.

4. **Molecular Polarity**
   Molecules that have polar covalent bonds and asymmetric electronic geometries have higher regions of electron density at one end of the molecule. Such molecules possess partial negative and positive charges and are called polar.

INSIGHT:
To count regions of high electron density simply include all bonding and lone pairs of electrons that surround the central atom. For example, H_2O has two bonding pairs and two lone pairs of electrons on the O atom. That amounts to four regions of high electron density.

YIELD

Electronic Geometry versus Molecular Geometry
All of the molecules described in this module have an electronic and a molecular geometry.
1) **Electronic geometry** considers all of the regions of high electron density including bonding pairs, lone pairs, and double or triple bonds.
2) **Molecular geometry** only considers those electrons and atoms that are involved in bonding pairs or in double and triple bonds.
The molecular geometry is different from the electronic geometry only in molecules having lone pairs of electrons.

Electron Groups on Central Atom	Electronic Geometry*			
	Orientation of Electron Groups	Description; Angles†	Line Drawing‡	Ball and Stick Model
2		linear; 180°		
3		trigonal planar; 120°		
4		tetrahedral; 109.5°		

Linear Molecules Sample Exercise

1. *What are the correct molecular geometries of these molecules?*
   BeI₂, BeHF

   The correct answer is linear for both molecules.

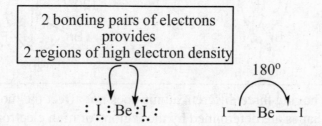

2 bonding pairs of electrons provides
2 regions of high electron density

This molecule's linear shape is determined by the electrons surrounding the central Be atom, not by the lone pairs on the I atoms.

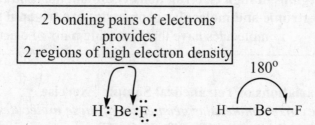

2 bonding pairs of electrons provides
2 regions of high electron density

The linear shape is determined by the 2 regions of high electron density on the central Be atom and is unaffected by the two different atoms bonded to Be.

INSIGHT: Covalent Compounds of Be do not obey the octet rule. If Be is the central atom in a molecule there are **two regions of high electron density** giving **linear electronic and molecular geometries**.

87

Trigonal Planar Molecules Sample Exercise

2. *What are the correct molecular geometries of these molecules?*

BH_3, AlHFBr

The correct answer is trigonal planar for both molecules.

3 bonding pairs of electrons give 3 regions of high electron density

$120°$ $\overset{H}{\underset{H \qquad H}{|}}$ $120°$

$120°$

3 bonding pairs of electrons give 3 regions of high electron density

$120°$ $\overset{H}{\underset{:Br \qquad F}{|}}$ $120°$

$120°$

Again, having three different atoms does not affect the molecule's shape. Shapes are determined by the regions of high electron density.

INSIGHT: Covalent compounds of the IIIA group (B, Al, Ga, & In) do not obey the octet rule. If a IIIA element is the central atom, **the molecule has three regions of high electron density** around the central atom and the **electronic and molecular geometries are trigonal planar**. These molecules have three bonding pairs of electrons.

Tetrahedral and Variations of Tetrahedral Sample Exercise

3. *What are the correct molecular geometries of these molecules?*

SiH_4, PF_3, H_2O, HBr

The correct answer is tetrahedral for SiH_4, trigonal pyramidal for PH_3, bent or angular for H_2O, and linear for HBr.

4 bonding pairs provide 4 regions of electron density

bonding pairs

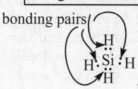

| 109.5° $\quad$ 109.5°
 109.5° H 109.5° |

3 bonding pairs + 1 lone pair provide 4 regions of electron density	Lone pairs occupy larger volumes than bonding pairs.

lone pair

$$H \cdot \overset{..}{\underset{H}{P}} \cdot H$$

107° H 107°

The lone pair is possible because P has one more valence electron than Si.	The lone pair crowds the bonding pairs decreasing their bond angles.

2 bonding pairs + 2 lone pairs provide 4 regions of electron density	2 lone pairs occupy an even larger volume than one lone pair.

2 lone pairs

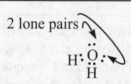

104.5° H

O has two more valence electrons than Si and can have 2 lone pairs.	The 2 lone pairs crowd the bonding pairs decreasing the bond angles more than in PH_3.

1 bonding pair + 3 lone pairs provide 4 regions of electron density	3 lone pairs surround the Br atom.

3 lone pairs

$$H : \overset{..}{\underset{..}{Br}} :$$

180°

H —— Br

Br has 3 more valence electrons than Si and can have 3 lone pairs.	There are only two atoms in this molecule, H and Br. The only possible geometry is linear.

All molecules in this category obey the octet rule having four regions of high electron density composed of eight electrons.

1. If a **IVA** element (C, Si, or Ge) is the central atom, the **electronic** and **molecular geometries are tetrahedral**. These molecules contain four bonding pairs of electrons.

2. If a **VA** element (N, P, or As) is the central atom, the **electronic geometry is tetrahedral** and the **molecular geometry is trigonal pyramidal**. These molecules contain three bonding pairs and 1 lone pair of electrons.

3. If a **VIA** element (O, S, Se) is the central atom, the **electronic geometry is tetrahedral** and the **molecular geometry is bent, angular, or V-shaped**. These molecules contain two bonding pairs and two lone pairs of electrons.

4. If a **VIIA** element (F, Cl, Br, or I) is the central atom, the **electronic geometry is tetrahedral** and the **molecular geometry is linear**. These molecules contain one bonding pair and three lone pairs of electrons.

Polarity of Molecules Sample Exercise

4. *Which of the following molecules are polar?*

$$CO_2, BH_3, CH_2F_2, H_2O$$

The correct answer is CO_2 and BH_3 are nonpolar while CH_2F_2 and OCl_2 are polar.

In Module 12 we discussed the concept of polar covalent bonds, covalent bonds in which the two atoms have different electronegativities. As a consequence, electrons in a polar bond are shared unequally. An entire molecule can also be polar or nonpolar. *Molecules that contain polar bonds are not necessarily polar*. To be polar a molecule must meet two necessary conditions.

1) The molecule must contain <u>at least</u> one polar bond or one lone pair of electrons.

2) The molecular geometry must be asymmetrical so that the partial negative or positive charges of the polar bonds do not cancel each other.

Let's use our problem solving steps check list to work the CO_2 problem.

1. **Read the problem**
   a) We are being asked to determine if CO_2 is a polar or nonpolar molecule.
2. **Understand precisely what the problem is asking**
   a) We must decide if CO_2 is polar or nonpolar.
3. **Identify the chemical principles**
   a) While the major principle is molecular polarity there are some other principles involved as well.
   1) Draw the correct Lewis dot structure for CO_2.
   2) Determine if the molecule has polar covalent bonds.

90

3) Determine the correct molecular shape.

4) Determine if the polar covalent bonds are symmetrically or asymmetrically arranged so that they cancel or reinforce the partial negative or positive charges of the polar covalent bonds.

4. Determine the relevant pieces of information

a) The most important pieces of information are the words polar and the molecular formula CO_2.

5. Create a solution pathway

a) Draw the correct Lewis dot structure of CO_2.

1) $N = 8$ (for C) $+ 2\times8$ (for O) $= 24$

2) $A = 4$ (for C) $+ 2\times6$ (for O) $= 16$

3) $S = 24 - 16 = 8$ electrons in bonding pairs

4) $A - S = 16 - 8 = 8$ electrons in lone pairs

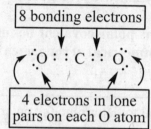

5) To help visualize this molecule use a line drawing

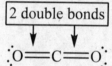

b) Does the molecule contain polar bonds?

1) C and O have different electronegativities.

2) O which is farther to the right on the periodic table is more electronegative than C.

3) Each of the double bonds in this molecule is a polar covalent bond.

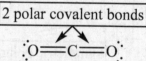

c) What is the molecular shape of CO_2?

1) Because each double bond counts as one region of high electron density on the central C atom, CO_2 has two regions of high electron density.

2) Molecules with two regions of high electron density are linear.

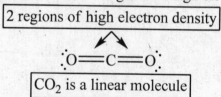

d) Are the polar covalent bonds arranged symmetrically or asymmetrically?
   1) A linear geometry is a symmetrical arrangement of the two polar covalent double bonds.

> O is more electronegative than C.
> Electrons in double bonds are
> **symmetrically** distributed about C atom

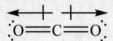

e) Because the polar covalent bonds are arranged symmetrically around the central C atom this is a **nonpolar molecule**.

6. **Check your answer**
   a) The molecule has a linear, symmetric arrangement of polar bonds on the central C atom.
   b) **Notice this molecule contains polar covalent bonds but the molecule is nonpolar because of the symmetrical arrangement of the polar bonds!**

INSIGHT:

> These same steps are used to determine the polarity of the following three molecules.

> BH_3, a trigonal planar molecule, contains 3 polar B-H bonds arranged **symmetrically** around the central B atom. The polar bonds cancel each other's effect and the molecule is nonpolar.

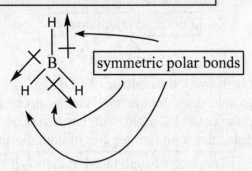

symmetric polar bonds

CH$_2$F$_2$, a terahedral molecule, contains 4 polar bonds. The two C-H bonds have their polar bond negative ends pointed toward the C atom (C is more electronegative than H). The two C-F bonds have their polar bond negative ends pointed away from the C atom (F is more electronegative than C). The result is an **asymmetrical** bond arrangement about the central C atom making the molecule polar.

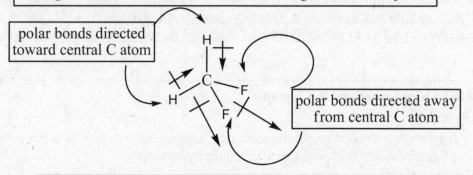

polar bonds directed toward central C atom

polar bonds directed away from central C atom

INSIGHT: Because CH$_2$F$_2$ is tetrahedral, every possible correct arrangement of the atoms in CH$_2$F$_2$ must be polar.

H$_2$O has two polar bonds and two lone pairs. Both negative ends of the polar O-H bonds are directed toward the O atom (O is more electronegative than H) reinforcing the large negative effect of the lone pairs making H$_2$O quite polar.

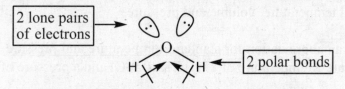

2 lone pairs of electrons

2 polar bonds

asymmetric arrangement of 2 polar bonds and 2 lone pairs around the central O atom

Module 14
Gases

Introduction

This module describes the basic laws that govern gases. In this module, the most important things to understand are <u>a) the kinetic-molecular theory of gases, b) how to use several gas laws including Boyle's, Charles', combined, Avogadro's, Dalton's, and the ideal gas law as well as c) the definition of STP and d) reaction stoichiometry using gas laws.</u>

Module 14 Key Equations & Concepts

1. $P_1V_1 = P_2V_2$

    Boyle's law is used to determine a new volume or pressure of a gas given the original volume and pressure at constant temperature.

2. $\dfrac{V_1}{T_1} = \dfrac{V_2}{T_2}$

    Charles' law is used to determine a new volume or temperature of a gas given the original volume and temperature at constant pressure.

3. $\dfrac{P_1V_1}{T_1} = \dfrac{P_2V_2}{T_2}$

    The combined gas law is a combination of Boyle's and Charles's gas laws. It is used to determine a new temperature, volume, or pressure of a gas given the original temperature, volume and pressure.

4. **STP**

    STP is an abbreviation for <u>standard temperature and pressure.</u> For gases these values are a temperature of 273.15 K (0.0°C) and a pressure of 1.00 atm.

5. $V = k\,n$

    Avogadro's law indicates that the volume of a gas in a container is directly related to the number of moles of the gas inside the container.

6. **Standard Molar Volume**

    One mole of any gas at STP has a volume of 22.4 L.

7. $PV = nRT$

    The ideal gas law is used to calculate any one of these quantities, the pressure, volume, temperature, or number of moles of a gas given three of the other quantities. This equation is often used in reaction stoichiometry problems involving gases.

8. $P_{total} = P_1 + P_2 + P_3 + \dots.$

    Dalton's law indicates that in mixtures of gases the total pressure of the gas mixture is the sum of the individual pressures of each gas in the mixture.

Kinetic Molecular Theory Sample Exercise

1. *In a comparison of equal mass samples of liquid and gaseous water both at $100^\circ C$, which of these are characteristics of gaseous water?*
   a. The water molecules are relatively close together.
   b. The water molecules are relatively far apart.
   c. The density is slightly less than 1.00 g/mL
   d. The density is much less than 1.00 g/mL.
   e. Attractive forces between the molecules are very weak.
   f. Attractive forces between the molecules are strong.

   Answers b, d, and e are characteristics of gaseous water.

INSIGHT: There are several important tenets of kinetic molecular theory (KMT) that you need to understand for this course. In KMT, gases are composed of small, independent particles that have little to no attractive forces between them and exist relatively far apart. These ideas explain the answer to exercise 1. Gaseous water molecules are quite far apart from one another. Liquid water has a density that is close to 1.00 g/mL because the water molecules are fairly close together. However, gas molecules are much farther apart, resulting in a density that is ~0.001 g/mL. Once you understand KMT, you will find that your mental picture of how atoms and molecules interact is vastly improved.

Boyle's Law Sample Exercise

2. *A sample of a gas initially having a pressure of 1.25 atm and volume of 3.50 L has its volume changed to 7.50 x 10^4 mL at constant temperature. What is the new pressure of the gas sample?*

   The correct answer is 0.0583 atm.

$$7.50 \times 10^4 \text{ mL} \left(\frac{1 \text{ L}}{1000 \text{ mL}} \right) = 75.0 \text{ L}$$

At constant T we can use Boyle's Law

$$P_1 V_1 = P_2 V_2 \text{ at constant temperature}$$

$$1.25 \text{ atm} \times 3.50 \text{ L} = P_2 \times 75.0 \text{ L}$$

$$\frac{1.25 \text{ atm} \times 3.50 \text{ L}}{75.0 \text{ L}} = P_2$$

$$0.0583 \text{ atm} = P_2$$

INSIGHT: It is very important in gas law problems that the proper units be used. In exercise 1, the gas volumes must be in the same units (either mL or L) to properly apply Boyle's law.

Charles' Law Sample Exercises

3. *A gas sample having a constant pressure of 1.75 atm and an initial volume of 4.50 L at 25.0°C is heated to 37.0°C. What is the gas's new volume?*

The correct answer is 4.68 L.

$$\frac{V_1}{T_1} = \frac{V_2}{T_2} \text{ where:}$$

$$V_1 = 4.50 \text{ L}, \ T_1 = 25.0°C = 298.1 \text{ K,}$$

$$\text{and } T_2 = 37.0°C = 310.1 \text{ K}$$

$$V_2 = \frac{V_1 T_2}{T_1} = \frac{(4.50 \text{ L}) \ (310.1 \text{ K})}{(298.1 \text{ K})} = 4.68 \text{ L}$$

 YIELD | All gas law problems involving temperature must be in units of Kelvin. Be absolutely certain that you convert temperatures into Kelvin when working any gas law problems.

Combined Gas Law Sample Exercise

4. *A gas sample initially having a pressure of 3.68 atm and a volume of 5.92 L at a temperature of 35.0°C is changed to STP. What is the gas's new volume?*

The correct answer is 24.6 L.

$$\frac{P_1 V_1}{T_1} = \frac{P_2 V_2}{T_2} \text{ where:}$$

$$P_1 = 3.68 \text{ atm}, \ V_1 = 5.92 \text{ L}, \ T_1 = 35.0°C = 308.1 \text{ K}$$

$$P_2 = 1.00 \text{ atm and } T_2 = 0.00°C = 273.1 \text{ K}$$

$$V_2 = \frac{P_1 V_1 T_2}{T_1 P_2} = \frac{(3.68 \text{ atm}) \ (5.92 \text{ L}) \ (308.1 \text{ K})}{(273.1 \text{ K}) \ (1.00 \text{ atm})} = 24.6 \text{ L}$$

INSIGHT: | STP is an abbreviation for standard temperature and pressure. When STP is used in a problem involving gases, you should use a temperature of 0.00° C or 273.15 K and pressure of 1.00 atm or 760 mm Hg.

Avogadro's Law Sample Exercise

5. *At 373.1 K and 2.00 atm, the Avogadro's law constant has a value of 15.3 L/mol. Determine the volume of 5.00 moles of $CO_2(g)$ at 373.1 K and 2.00 atm.*

The correct answer is 76.5 L.

$$V = k \, n$$

$$V = \left(15.3 \ \text{L}\big/\text{mol} \right) 5.00 \text{ mol} = 76.5 \text{ L}$$

96

The value of the Avogadro's law constant changes with temperature and pressure. At STP, it has a value of 22.4 L/mol. Consequently, if you know the volume of a gas at STP, it is easy to determine the number of moles or grams of the gas as shown in the next exercise.

Standard Molar Volume Sample Exercise

6. *How many grams of $CO_2(g)$ are present in 11.2 L of the gas at STP?*

The correct answer is 22.0 g

$$n = 11.2 \text{ L} \left(\frac{1 \text{ mol}}{22.4 \text{ L}}\right) = 0.500 \text{ mol}$$

$$0.500 \text{ mol} \left(\frac{44.0 \text{ g } CO_2}{1 \text{ mol } CO_2}\right) = 22.0 \text{ g } CO_2$$

Ideal Gas Law Sample Exercise

7. *A gas sample at a pressure of 3.50 atm and a temperature of 45.0°C has a volume of 1.65×10^3 mL. How many moles of gas are in this sample?*

The correct answer is 0.221 moles.

$$PV = nRT \text{ therefore } n = \frac{PV}{RT}$$

In this problem

$$P = 3.50 \text{ atm}, \quad V = 1.65 \times 10^3 \text{ mL} = 1.65 \text{ L},$$

$$R = 0.0821 \frac{L \text{ atm}}{\text{mol K}}, \quad T = 45.0°C = 318.1 \text{ K}$$

$$n = \frac{PV}{RT} = \frac{(3.50 \text{ atm})(1.65 \text{ L})}{\left(0.0821 \frac{L \text{ atm}}{\text{mol K}}\right)(318.1 \text{ K})} = 0.221 \text{ mol}$$

INSIGHT: R is the ideal gas constant. In gas laws, it has the following value and units, R = 0.0821 L atm/mol K. This defines the units that we must use in the ideal gas law. **P must be in atm, V in L, n in moles, and T in K**.

Dalton's Law Sample Exercise

8. *A gaseous mixture is composed of 0.10 atm of CO_2, 0.50 atm of N_2, 0.40 atm of O_2, and 0.20 atm of He. What is the total pressure of this mixture?*

The correct answer is 1.20 atm.

$$P_{total} = P_1 + P_2 + P_3 + P_4$$

$$P_{total} = 0.10 \text{ atm} + 0.50 \text{ atm} + 0.40 \text{ atm} + 0.20 \text{ atm}$$

$$P_{total} = 1.20 \text{ atm}$$

Reaction Stoichiometry Involving Gases

9. *If 35.0 g of Al are reacted with excess sulfuric acid, how many L of hydrogen gas, H_2, are formed at STP?*

$$2\ Al(s)\ +\ 3\ H_2SO_4(aq)\ \rightarrow\ Al_2(SO_4)_3(aq)\ +\ 3\ H_2(g)$$

The correct answer is 43.7 L

a) Calculate the number of moles of hydrogen gas formed in the reaction.

$$(35.0\ g\ Al)\left(\frac{1\ mol}{26.98\ g\ Al}\right)\left(\frac{3\ mol\ H_2}{2\ mol\ Al}\right)=1.95\ mol$$

b) Because the conditions are at STP, we can use the standard molar volume to determine the volume of H_2 produced.

$$V\ =\ 22.4\ L/mol\ (1.95\ mol)=43.7\ L$$

<YIELD> If the reaction conditions are at STP, using the standard molar volume is a simple method to determine either the volume or the number of moles. **If reaction conditions are not at STP, you cannot use this method**. The standard molar volume is 22.4 L only for gases, **not** for liquids or solids!

10. *If 105.0 g of Al are reacted with excess sulfuric acid, how many L of hydrogen gas, H_2, are formed at 1.25 atm of pressure and $75.0^\circ C$?*

$$2\ Al(s)\ +\ 3\ H_2SO_4(aq)\ \rightarrow\ Al_2(SO_4)_3(aq)\ +\ 3\ H_2(g)$$

The correct answer is 133 L.

This is the most encompassing problem encountered in this module using many of the concepts introduced. It is a good one to use the problem solving steps check list to help determine the final answer.

1. **Read the problem**
2. **Understand precisely what the problem is asking**
   a) We must determine the volume of H_2 gas formed by the reaction of Al with H_2SO_4 at 1.25 atm and $75.0^\circ C$.
3. **Identify the chemical principles**
   a) Reaction stoichiometry
   b) Ideal gas law
   c) Mass to mole conversions
4. **Determine the relevant pieces of information**
   a) The 105.0 g of Al is very important.
   b) The balanced chemical reaction is a necessary part of the solution pathway.
   c) The fact that H_2 is a gas indicates that we must use the gas laws.
   d) The pressure of 1.25 atm and temperature of $75.0^\circ C$ indicate that we cannot use the standard molar volume pathway but must use the ideal gas law.
5. **Create a solution pathway**
   a) Using reaction stoichiometry, calculate the number of moles of H_2 gas made in the reaction.

$$(105.0 \text{ g Al}) \left(\frac{1 \text{ mol}}{26.98 \text{ g Al}} \right) \left(\frac{3 \text{ mol H}_2}{2 \text{ mol Al}} \right) = 5.838 \text{ mol H}_2$$

b) Convert 75.0°C to K.

$$75.0^{\circ}\text{C} = 273.15 + 75.0^{\circ}\text{C} = 348.1 \text{ K}.$$

c) Use ideal gas law to determine the volume of H_2 gas at 1.25 atm and 348.1 K.

$$PV = nRT \text{ thus } V = \frac{nRT}{P}$$

$$V = \frac{(5.838 \text{ mol}) \left(0.0821 \text{ L} \cdot \text{atm} \middle/ \text{mol} \cdot \text{K} \right) (348.1 \text{ K})}{1.25 \text{ atm}} = 133 \text{ L}$$

6. **Check your answer**
   a) The number of moles of H_2 gas made in this reaction seems reasonable.
   b) We converted $^{\circ}$C to K.
   c) We have appropriately used the ideal gas law.
   d) Our volume of H_2 gas made seems quite reasonable, especially when compared to the answer in the previous exercise.
   e) Our answer is correct.

99

Module 15
Liquids and Solids

Introduction
In this module, we will look at some of the properties of liquids and solids. The most important things to understand are a) using the basic heat transfer equation to determine the amount of heat necessary to warm a substance that remains in a single phase, b) how to determine the amount of heat necessary to cause a substance to change phase, c) how to determine which intermolecular forces are the most significant in a substance, d) how to determine the relative freezing and boiling points of various liquids based on their intermolecular forces, and e) how to determine the relative melting points of various solids based on the strength of their bonding.

Module 15 Key Equations & Concepts

1. **$q = mC\Delta T$**
   This is the basic heat transfer equation to calculate the amount of energy emitted or absorbed when an object warms up or cools down. (q is the heat, m is the mass, C is the specific heat, and ΔT is the temperature change.)

2. **$q = m\,\Delta H$**
   This equation is used to determine the amount of heat necessary to convert a substance from one state of matter to another, for example from solid to liquid or liquid to gas.

3. **Dipole-dipole interactions, Hydrogen bonding, London Dispersion forces**
   These are the three intermolecular forces involved in liquids and solids. The strength of these interactions determines the boiling points of various liquids.

4. **Covalent network solids, Ionic solids, Metallic solids, Molecular solids**
   These are the four classifications of solids. The strength of the intermolecular forces or bonds in solids determines their freezing and boiling points.

Kinetic Molecular Theory of Liquids and Solids Sample Exercise

1. *Determine which of these statements are kinetic molecular theory (KMT) descriptions of solids and which are descriptions of liquids.*

   a. *Individual particles are in a close but random motion relative to one another.*

   b. *Individual particles are in fixed positions relative to one another.*

   c. *In general, the substance's density is the highest of the three common states of matter.*

   d. *In general, the substance's density is intermediate of the three common states of matter.*

   e. *Attractions between molecules, atoms, or ions of the substance are very strong.*

   f. *Attractions between molecules, atoms, or ions of the substance are intermediate in strength.*

   The correct answer is a, d, and f are KMT descriptions of liquids and b, c, and e are KMT descriptions of solids.

Heat Transfer Sample Exercises

2. How much heat is required to heat 75.0 g of aluminum from 25.0°C to 175.0°C? The specific heat of aluminum is 0.900 J/g °C.

The correct answer is 1.01×10^4 J or 10.1 kJ.

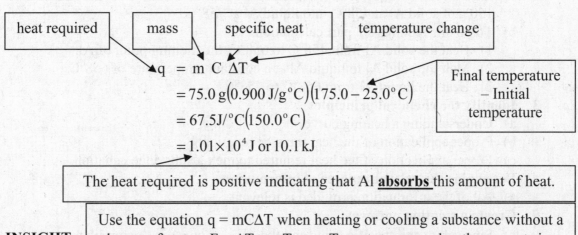

| heat required | mass | specific heat | temperature change |

$$q = m \ C \ \Delta T$$
$$= 75.0 \, g(0.900 \, J/g°C)(175.0 - 25.0°C)$$

Final temperature – Initial temperature

$$= 67.5 J/°C(150.0°C)$$
$$= 1.01 \times 10^4 J \text{ or } 10.1 \, kJ$$

The heat required is positive indicating that Al **absorbs** this amount of heat.

INSIGHT: Use the equation $q = mC\Delta T$ when heating or cooling a substance without a change of state. For ΔT use $T_{Final} - T_{Initial}$ to ensure q has the correct sign.

3. How much heat is required to convert 100.0 g of solid Al at 658°C into liquid Al at 658°C? The melting point of Al is 658°C. The ΔH_{fusion} for Al = 10.6 kJ/mol.

The correct answer is q = 39.3 kJ.

Because the ΔH_{fusion} is in units of kJ/mol, we must convert g of Al to mol of Al.

$$? \, mol \, Al = 100.0 \, g \, Al \left(\frac{1 \, mol \, Al}{26.98 \, g \, Al} \right) = 3.706 \, mol \, Al$$

$$q = m\Delta H_{fusion} = 3.706 \, mol \, Al \left(\frac{10.6 \, kJ}{mol \, Al} \right) = 39.3 \, kJ$$

| mol of Al | ΔH_{fusion} for Al |

INSIGHT: Use the equation $q = m\Delta H$ when a substance changes state from solid to liquid, liquid to gas, solid to gas, etc. ΔH_{fusion} is used when a solid changes to a liquid. $\Delta H_{vaporization}$ is used when a liquid changes to a gas. $\Delta H_{sublimation}$ is used when a solid changes to a gas.

4. *How much heat is required to convert 150.0 g of solid Al at 458°C into liquid Al at 758°C? The melting point of Al is 658°C. The specific heats for Al are, C_{solid} = 24.3 J/mol °C and C_{liquid} = 29.3 J/mol °C. The ΔH_{fusion} for Al = 10.6 kJ/mol.*
The correct answer is q = 102.2 kJ.

This is a relatively complex problem requiring several steps with several places where mistakes can be easily made. Let's use the problem solving steps check list to work this problem.

1. **Read the problem**
   a) This problem has an enormous amount of information for you to compile. Recognize that much of it is data that you must use in the number solving part of the problem.
2. **Understand precisely what the problem is asking**
   a) This problem is asking us to calculate the amount of heat necessary to convert 150 g of solid Al at 458°C into liquid Al at 758°C.
   b) To do this requires a 3 part calculation.
      1) Heat the solid Al from 458°C to 658°C, the melting point of Al.
      2) Melt the solid Al to liquid Al at a constant temperature of 658°C.
      3) Heat the liquid Al from 658°C to 758°C.
3. **Identify the chemical principles**
   a) Understanding a heating curve.
   b) Proper application of the heat transfer equation
   c) Proper application of the heat required to melt a substance equation
4. **Determine the relevant pieces of information**
   a) All of the information provided is relevant.
5. **Create a solution pathway**
   a) Since all of the specific heats and the ΔH_{fusion} given in this problem are in J or kJ/mol, we must convert the mass of Al to moles of Al.

$$150.0 \text{ g} \left(\frac{1 \text{ mol Al}}{26.98 \text{ g}} \right) = 5.560 \text{ mol Al}$$

   b) Look at the heating curve for Al given below. In this step we will determine the heat necessary to warm solid Al from 458°C to its melting point, 658°C.

$$q = mC_{solid}\Delta T = 5.560 \text{ mol} \left(24.3 \text{ J/mol °C} \right) \left(658°C - 458°C \right)$$

$$= 135 \text{ J/ °C} \left(200°C \right)$$

$$= 2.70 \times 10^4 \text{ J} = 27.0 \text{ kJ}$$

Notice that in this step we use the specific heat for solid Al.	Solid Al is heated from 458°C to 658°C, $T_{final} - T_{initial}$.

   c) Look at the heating curve for Al given below. In this step we will determine the heat necessary to melt solid Al to liquid Al at its melting point, 658°C.

$$q = m\Delta H_{fusion}$$

$$= 5.560 \text{ mol} (10.6 \text{ kJ/mol}) = 58.9 \text{ kJ}$$

Notice that in this step we use the ΔH_{fusion} for Al. There is no T change.

102

d) Look at the heating curve for Al given below. In this step we will determine the heat necessary to warm liquid Al from 658°C to 758°C.

$$q = mC_{liquid}\Delta T = 5.560 \text{ mol}\left(29.3 \text{ J/mol °C}\right)\left(758°C - 658°C\right)$$

$$= 163 \text{ J/°C}\left(100°C\right)$$

$$= 1.63 \times 10^4 \text{ J} = 16.3 \text{ kJ}$$

Notice that in this step we use the specific heat for liquid Al.	Liquid Al is heated from 658°C to 758°C, $T_{final} - T_{initial}$.

e) The total amount of heat required to heat 150.0 g of solid Al from 458°C to its melting point of 658°C, melt it, then heat the 150.0 g of liquid Al from 658°C to 758°C is the sum of the three steps for this problem.

$$q_{Total} = 27.0 \text{ kJ} + 58.9 \text{ kJ} + 16.3 \text{ kJ} = 102.2 \text{ kJ}$$

6. Check your answer
a) We have correctly used the heating curve to determine the applicable steps.
b) We have correctly calculated the heat necessary to warm the solid Al.
c) We have correctly calculated the heat necessary to melt the solid Al.
d) We have correctly calculated the heat necessary to warm the liquid Al.
e) Our answer is correct.

Heating Curve for Al

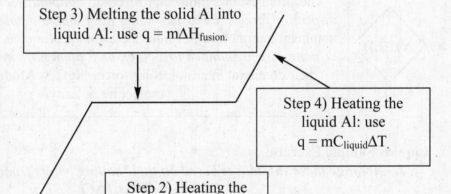

Sample exercise 4 involves only one phase change, namely converting solid Al into liquid Al. If a second phase change were included, converting solid Al into gaseous Al, the following steps would have to be included.

1) Heating the liquid Al to the boiling point using $q = mC_{liquid}\Delta T$.
2) Boiling the liquid Al using $q = m\Delta H_{vaporization}$.
3) Heating the gaseous Al using $q = mC_{gas}\Delta T$.

Intermolecular Forces Sample Exercises

5. *Arrange these intermolecular forces based on their strengths from lowest to highest.*

   Hydrogen bonding, London dispersion forces, Dipole-dipole forces

   The correct answer is London dispersion forces < Dipole-dipole forces < Hydrogen bonding.

6. *Determine the strongest intermolecular force present in each of these molecules.*

   CO_2, C_2H_5OH, CH_3Cl

   The correct answer is London dispersion forces for nonpolar CO_2, hydrogen bonding for polar C_2H_5OH, and dipole-dipole forces for polar CH_3Cl.

The strongest intermolecular forces in **nonpolar** *molecules are London dispersion forces*. **Polar** *molecules have dipole-dipole interactions as their dominant intermolecular force. Polar molecules containing one or more H atoms* **directly bonded to a N, O, or F atom** *have hydrogen bonding as their dominant intermolecular force*. Review Module 13 for rules on molecular polarity.

Liquids Sample Exercise

7. *Arrange these substances based on their boiling points from lowest to highest.*
   CO_2, C_2H_5OH, CH_3Cl

   The correct answer is $CO_2 < CH_3Cl < C_2H_5OH$.

Boiling points are determined by the strength of the intermolecular forces present in a liquid. In general, hydrogen bonding is strongest, dipole-dipole interactions are next strongest, and the weakest intermolecular force is London dispersion forces. The strongest intermolecular forces in liquid CO_2 are London dispersion forces, CH_3Cl's strongest intermolecular forces are dipole-dipole interactions, and hydrogen bonding is dominant in C_2H_5OH. Thus the correct order is $CO_2 < CH_3Cl < C_2H_5OH$.

Solids Sample Exercises

8. *Arrange, from lowest to highest, these solid types based on the strength of the forces which hold each solid together.*

     Molecular solids, Covalent network solids, Ionic solids, Metallic solids

     The correct answer is Molecular solids < Metallic solids < Ionic solids < Covalent network solids.

9. *Classify each of these solids into one of the four types of solids.*

     CO_2, KCl, Na, SiO_2

     The correct answer is CO_2 is a molecular solid, KCl is an ionic solid, Na is a metallic solid, and SiO_2 is a covalent network solid.

---

1) **Molecular solids** are covalent compounds containing individual molecules in the solid state. Molecular solids are held together by intermolecular forces. Most of the covalent species that you have learned up to now are molecular solids.

2) **Metallic solids** are by far the easiest to determine. Simply look for a metallic element. Metallic solids are held together with metallic bonds which are described by either the electron sea model or by band theory.

3) **Ionic solids** are the basic ionic compounds that you have learned up to this point. In the solid state, ionic solids consist of arrays of negative and positive ions attracted to one another by ionic bonds.

4) For most students, the hardest substances to classify are **network covalent solids**. These substances are covalent species that form large extended arrays of atoms or molecules held together by covalently bonds. Your instructor and textbook probably have a list of the common network covalent solids which they would like for you to know. Some common covalent network solids include diamond, graphite, tungsten carbide (WC), and sand (SiO_2).

---

10. *Arrange these substances based on their melting points from lowest to highest.*

     CO_2, KCl, Na, SiO_2

     The correct answer is CO_2 < Na < KCl < SiO_2.

INSIGHT: Melting points of solids are determined by the strength of the forces bonding them together. In general, the weakest forces are intermolecular forces found in molecular solids like CO_2, next weakest are metallic bonds as in Na, ionic bonds are relatively strong as in KCl, and the strongest forces are the atom to atom covalent bonds in network covalent solids like SiO_2.

Module 16
Solutions

Introduction

This module discusses the properties of solutions. The most important things to understand from the module are how to a) predict which species will dissolve in a given solvent based on molecular polarity, b) increase the solubility of a given species in a solvent, c) convert from one concentration unit to another, d) perform dilution and titration calculations, and e) determine the freezing and boiling points of solutions. All of these techniques are used for a variety of problems concerning solutions.

Module 16 Key Equations & Concepts

1. **Like Dissolves Like**

   Polar molecules are readily soluble in other polar molecules and that nonpolar molecules are readily soluble in other nonpolar molecules. However, polar molecules are fairly insoluble in nonpolar molecules.

2. **Solute solubility is increased when the solvent is *heated* in an *endothermic* dissolution. Solute solubility is increased when the solvent is *cooled* in an *exothermic* dissolution. Increasing a gas's pressure increases the solubility of a gas in a liquid.**

   These basic rules for changing the solubility of various substances in solution are used to determine methods to increase the solubility of substances.

3. $$M = \frac{\text{moles of solute}}{\text{L of solution}}, m = \frac{\text{moles of solute}}{\text{kg of solvent}}$$

   $$\% \text{ w/w} = \frac{\text{mass of one solution component}}{\text{mass of total solution}} \times 100$$

   Three common units of solution concentration are molarity (M) used in reaction stoichiometry, molality (m) used in the freezing point depression and boiling point elevation relationships, and percent by mass (% w/w) used frequently for concentrated solutions.

4. $M_1 V_1 = M_2 V_2$

   The dilution formula is used to calculate the amount of concentrated solution necessary to make a dilute solution.

5. **Titrations**

   One form of solution reaction stoichiometry are titrations. Titrations are frequently used to standardize solutions then determine the concentration of a second solution. Acid-base neutralization reactions are commonly used in titrations.

6. $\Delta T_f = K_f \times m$ and $\Delta T_b = K_b \times m$

   The freezing point depression and boiling point elevation relationships describe how much the freezing or boiling temperatures of a solution differ from the pure solvent's freezing and boiling points.

Solubility of a Solute in a Given Solvent Sample Exercise

1. *Which of the following substances are soluble in water?*

   CCl_4, NH_3, C_8H_{18}, $CaCl_2$, CH_3OH, $Ca_3(PO_4)_2$

   The correct answer is NH_3, $CaCl_2$, and CH_3OH are soluble in water. The others are insoluble in water.

INSIGHT:	The "Like Dissolves Like" rule implies that polar species dissolve in polar species and nonpolar species dissolve in nonpolar species. Consequently, nonpolar species do not dissolve in polar species and polar species do not dissolve in nonpolar species. In this example, NH_3 and CH_3OH are both polar covalent compounds which dissolve in the highly polar solvent water. $CaCl_2$ is an ionic compound which is water soluble (the solubility rules also apply in these problems). CCl_4 and C_8H_{18} are both nonpolar covalent compounds thus they are insoluble in water. $Ca_3(PO_4)_2$ is an ionic compound that is insoluble in water. Refresh your memory on the solubility guidelines if necessary.
YIELD	In these problems there are two important considerations to have in mind. 1) Are the covalent compounds in the problem polar or nonpolar? 2) Are the ionic compounds in the problem soluble or insoluble based on the solubility rules? Remember the strong acids and bases from previous modules are also water soluble.

Molecular Representation of Solubility Sample Exercise

2. *Choose the diagram from below which best represents at the molecular level an ionic solid dissolved in water.*

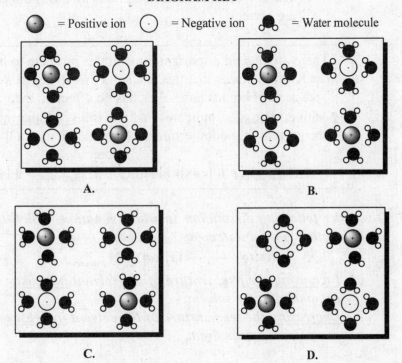

107

The correct answer is diagram D.

<table>
<tr><td>INSIGHT:</td><td>When an ionic solid dissolves in water, water molecules surround the ions. Dissolution occurs because the polar water molecules are attracted by the charged ions. The partially negative O atoms on the water molecule get closest to the positive ions. The partially positive H atoms are attracted to the negative ions. Only diagram D. has the ions surrounded correctly.</td></tr>
</table>

Increasing the Solubility of a Solute in a Given Solvent Sample Exercises

3. *Given the following dissolution in water equation, which of these changes in conditions are correct statements?*

$$KI(s) \xrightarrow{H_2O} K^+(aq) + I^-(aq) \quad \Delta H_{dissolution} > 0$$

a) *<u>Increasing</u> the temperature of the solvent <u>increases</u> the solubility of the compound in the solvent.*

b) *<u>Decreasing</u> the temperature of the solvent <u>increases</u> the solubility of the compound in the solvent.*

c) *Changing the temperature of the solvent does <u>not affect</u> the solubility of the compound in the solvent.*

d) *<u>Increasing</u> the pressure of the solute <u>increases</u> the solubility of the compound in the solvent.*

e) *<u>Increasing</u> the pressure of the solute <u>decreases</u> the solubility of the compound in the solvent.*

f) *<u>Increasing</u> the pressure of the solute does <u>not affect</u> the solubility of the compound in the solvent.*

The correct answer is conditions a) and f) are correct and no others are correct.

<table>
<tr><td>INSIGHT:</td><td>There are several important hints in this problem to help you answer it. The KI(s) indicates that this is a solid dissolving in water. Changing the pressure of liquids and solids has no effect on their solubilities. The positive $\Delta H_{dissolution}$ indicates that this is an <u>endothermic</u> process. Heating the solvent for endothermic dissolutions increases the solubility of the solute.

$\Delta H_{dissolution} < 0$ is exothermic. $\Delta H_{dissolution} > 0$ is endothermic.</td></tr>
</table>

4. *Given the following dissolution in water equation, which of these changes in conditions are correct statements?*

$$O_2(g) \xrightarrow{H_2O} O_2(aq) \quad \Delta H_{dissolution} < 0$$

a) *<u>Increasing</u> the temperature of the solvent <u>increases</u> the solubility of the compound in the solvent.*

b) *<u>Decreasing</u> the temperature of the solvent <u>increases</u> the solubility of the compound in the solvent.*

c) *Changing the temperature of the solvent does <u>not affect</u> the solubility of the compound in the solvent.*

d) *<u>Increasing</u> the pressure of the solute <u>increases</u> the solubility of the compound in the solvent.*

e) *<u>Increasing</u> the pressure of the solute <u>decreases</u> the solubility of the compound in the solvent.*

f) *<u>Increasing</u> the pressure of the solute does <u>not affect</u> the solubility of the compound in the solvent.*

The correct answer is conditions b) and d) are correct and no others are correct.

INSIGHT: The important hints in this problem are 1) $O_2(g)$ indicates that this is a gas dissolving in water. Increasing the pressure of gases has a significant effect on their solubilities. In general, increasing the pressure of a gas increases its solubility in a liquid. 2) The negative $\Delta H_{dissolution}$ indicates that this is an <u>exothermic</u> process. Heating the solvent for exothermic dissolutions decreases the solubility of the solute. Cooling the solvent increases the solubility of the solute in exothermic dissolutions.

 YIELD 1) Pay attention to whether the substance being dissolved is a solid, liquid, or gas. Gases are more soluble at high pressure but solids and liquids are not. That will tell you if the changing pressure condition is applicable. 2) Pay attention to whether or not the dissolution is endo- or exothermic. That is a hint as to whether heating or cooling the solvent increases the solubility of the substance.

Concentration Unit Sample Exercises

5. *An aqueous hydrochloric acid solution contains 9.13 g of HCl in every 100 mL of solution. What is the concentration of this solution in molarity (M)?*

The correct answer is 2.50 M.

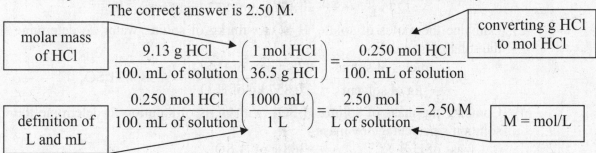

6. *An aqueous sulfuric acid solution that is 3.75 M has a density of 1.225 g/mL. What is the concentration of this solution in molality (m) and percent by mass (%) of H_2SO_4?*

The correct answer is 4.38 *m* and 30.0 % w/w.

This problem is relatively complex. Let's use the problem solving steps check list to outline the procedure for working these types of problems.

1. **Read the problem**
2. **Understand precisely what the problem is asking**
   a) We are to take a solution concentration in molarity and convert it to molality and percent by mass.
3. **Identify the chemical principles**
   a) Concentration units molarity (M), molality (m), and percentage by mass (%)
   b) Differences between solute, solvent, and solution
   c) Correct use of density
4. **Determine the relevant pieces of information**
   a) All the parts of this problem are relevant.
5. **Create a solution pathway**
   a) Recognize that molarity is defined as moles of solute/L of solution. Because the denominator in the molarity fraction is L of solution (which has both the solute and solvent intertwined) our first step in this process is to separate the solute and solvent so that we know how much of each is present.

$$3.75 \text{ M } H_2SO_4 = \frac{3.75 \text{ moles of } H_2SO_4}{1.00 \text{ L of solution}}$$

   b) In concentration conversion problems, we can start with any amount of solution. It is a good idea to use an amount that makes solving this problem as simple as possible. In this case, that is 1.00 L of solution. Use the solution's density to determine the mass of the solution.

$$1.00 \text{ L of } 3.75 \text{ M } H_2SO_4 = 1000 \text{ mL} \left(\frac{1.225 \text{ g}}{\text{mL}} \right) = 1225 \text{ g of } 3.75 \text{ M } H_2SO_4$$

   c) From the molar mass of H_2SO_4 and the 3.75 M, determine the mass of solute, H_2SO_4, in the solution.

$$3.75 \text{ mole } H_2SO_4 \left(\frac{98.1 \text{ g } H_2SO_4}{1 \text{ mole } H_2SO_4} \right) = 368 \text{ g } H_2SO_4$$

   d) Use the mass of the solution and the mass of the solute to determine the mass of the solvent water in 1.00 L of solution.

$$1225 \text{ g} - 368 \text{ g} = 857 \text{ g or } 0.857 \text{ kg of water}$$

   e) Combine the moles of solute, H_2SO_4, with kg of solute, water, to determine the solution's molality.

$$m = \frac{\text{moles of solute}}{\text{kg of solvent}} = \frac{3.75 \text{ moles of } H_2SO_4}{0.857 \text{ kg of } H_2O} = 4.38 \text{ } m \text{ } H_2SO_4$$

   f) Combine the mass of solute, H_2SO_4, with mass of solution to determine the solution's percentage by mass.

$$\% = \frac{\text{mass of } H_2SO_4}{\text{mass of solution}} \times 100\% = \frac{368 \text{ g of } H_2SO_4}{1225 \text{ g of solution}} \times 100\% = 30.0\% \text{ w/w } H_2SO_4$$

6. **Check your answer**
   a) If the solute is water, the molarity and molality of a solution have similar values as is the case for this solution, 3.75 M and 4.38 m.
   b) Once the masses of the solvent and solute are separated, determining the % by mass is straightforward.
   c) Our answers are correct.

110

Converting solution concentrations from molarity to the other two concentration units is by far the hardest solution conversion problem. The key to accomplishing this is separating the masses of the solute and solvent from the mass of the solution.

7. *A solution is made of 25.0 g of phenol dissolved in 75.0 g of benzene. What is the concentration of this solution in percentage by mass (% w/w) of phenol?*

The correct answer is 25.0% w/w.

Total mass of solution = 25.0 g phenol + 75.0 g benzene = 100.0 g solution

$$\% \text{ phenol} = \frac{25.0 \text{ g solute}}{100.0 \text{ g solution}} \times 100\% = 25.0\% \text{ w/w phenol}$$

Determining the mass of the solution.

This solution is also 75.0% w/w benzene.

8. *An aqueous sucrose, $C_{12}H_{22}O_{11}$, solution that is 11.0 % by mass has a density of 1.0432 g/mL. What is the concentration of this solution in molarity (M) and molality (m)?*

The correct answer is 0.335 M and 0.361 *m*.

By assuming that we have 100.0 g of this sucrose solution, we immediately know that there are 11.0 g of sucrose (11% of 100 g) and 89.0 g of water (100% - 11% = 89%). In one simple step, we have separated the masses of solute and solvent which is the key to solving all of the concentration conversion problems.

a) 11.0 g of $C_{12}H_{22}O_{11} \left(\dfrac{1 \text{ mole of } C_{12}H_{22}O_{11}}{342.3 \text{ g of } C_{12}H_{22}O_{11}} \right) = 0.0321$ moles of $C_{12}H_{22}O_{11}$, the solute

b) volume of 100.0 g of this solution $= 100.0 \text{ g} \left(\dfrac{1.00 \text{ mL}}{1.0432 \text{ g}} \right) = 95.9 \text{ mL} = 0.0959 \text{ L}$

Converting the solution mass to volume.

c) M $= \dfrac{\text{moles of sucrose}}{\text{L of solution}} = \dfrac{0.0321 \text{ moles of sucrose}}{0.0959 \text{ L of solution}} = 0.335 \, M$

d) $m = \dfrac{\text{moles of sucrose}}{\text{kg of solvent}} = \dfrac{0.0321 \text{ moles of sucrose}}{0.0890 \text{ kg of water}} = 0.361 \, m$

Concentrations are easily determined from masses, moles, and volume.

INSIGHT: Notice that this problem is much easier than exercise 6 because % by mass is a concentration unit that easily separates into the solute and the solvent.

9. *An aqueous NaCl, solution consists of 11.0 g of NaCl dissolved in 89.0 g of water. What is the concentration of this solution in molality (m)?*

The correct answer is 2.11 *m*.

Determine moles of solute.

111

$$? \text{ mol NaCl} = 11.0 \text{ g NaCl}\left(\frac{1 \text{ mol NaCl}}{58.44 \text{ g NaCl}}\right) = 0.188 \text{ mol}$$

$$? \text{ kg water} = 89.0 \text{ g water}\left(\frac{1 \text{ kg}}{1000 \text{ g}}\right) = 0.0890 \text{ kg}$$

Determine mass of solvent, in kg.

$$? \, m = \frac{0.188 \text{ mol solute NaCl}}{0.0890 \text{ kg solvent water}} = 2.11 \, m$$

m = mass of solute/kg of solvent.

Dilution Sample Exercise

10. How many mL of 12.0 M HCl are required to make 500.0 mL of 1.25 M HCl?
The correct answer is 52.1 mL.

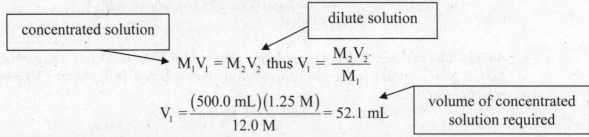

concentrated solution

dilute solution

$$M_1 V_1 = M_2 V_2 \text{ thus } V_1 = \frac{M_2 V_2}{M_1}$$

$$V_1 = \frac{(500.0 \text{ mL})(1.25 \text{ M})}{12.0 \text{ M}} = 52.1 \text{ mL}$$

volume of concentrated solution required

Titration Sample Exercise

11. How many mL of 0.125 M HCl are required to exactly neutralize 25.0 mL of an aqueous 0.025 M Sr(OH)$_2$ solution?
The correct answer is 10.0 mL.

INSIGHT: The words "exactly neutralize" or "neutralize" are your clue that this is an acid-base titration problem. You should also note that it is the reaction of a strong acid, HCl, with the dihydroxy strong base, Sr(OH)$_2$. In all titrations, the 1st step is to **write a balanced chemical reaction**.

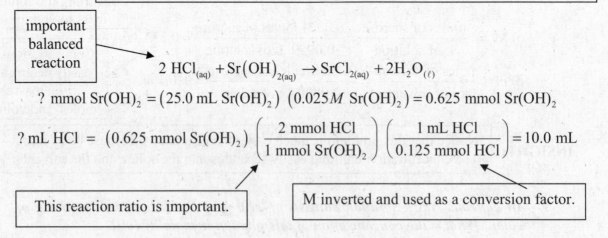

important balanced reaction

$$2 \text{ HCl}_{(aq)} + \text{Sr(OH)}_{2(aq)} \rightarrow \text{SrCl}_{2(aq)} + 2\text{H}_2\text{O}_{(\ell)}$$

$$? \text{ mmol Sr(OH)}_2 = (25.0 \text{ mL Sr(OH)}_2) \ (0.025 M \text{ Sr(OH)}_2) = 0.625 \text{ mmol Sr(OH)}_2$$

$$? \text{ mL HCl} = (0.625 \text{ mmol Sr(OH)}_2)\left(\frac{2 \text{ mmol HCl}}{1 \text{ mmol Sr(OH)}_2}\right)\left(\frac{1 \text{ mL HCl}}{0.125 \text{ mmol HCl}}\right) = 10.0 \text{ mL}$$

This reaction ratio is important.

M inverted and used as a conversion factor.

YIELD

A very common mistake made by many students is to use the dilution equation ($M_1V_1 = M_2V_2$) for titrations and other solution reaction problems. Compare exercises 10 and 11. They have some significant differences. 10 involves only one chemical species, HCl. Contrastingly, 11 involves HCl, $Sr(OH)_2$ and the products. Furthermore, 11 has a chemical reaction as a part of the problem. These are the clues that indicate 10 is a dilution problem, where $M_1V_1 = M_2V_2$ is applicable, and 11 is a titration.

Freezing Point Depression and Boiling Point Elevation Sample Exercises

12. *If 11.0 g of sucrose are dissolved in 89.0 g of water, at what temperature will this solution boil at 1.00 atm of pressure? The boiling point elevation constant, K_b, for water is 0.512 $^\circ C/m$.*

The correct answer is 100.185 $^\circ$C.

This solution's concentration was determined in exercise 8 to be 0.361 m.

$$\Delta T_b = K_b m = \left(0.512 \ ^\circ C/m\right)\left(0.361 \ m\right) = 0.185 \ ^\circ C$$

boiling point of the solution = 100.000 $^\circ$C + 0.185 $^\circ$C = 100.185 $^\circ$C

boiling point of pure water

boiling point increase due to sucrose

13. *A solution is made of 12.4 g of a compound dissolved in 100.0 g of water then the solution is frozen and found to have a freezing point of -5.00°C. What is the molar mass of the compound? The freezing point depression constant, K_f, for water is 1.86°C/m.*

The correct answer is 46.1 g/mol.

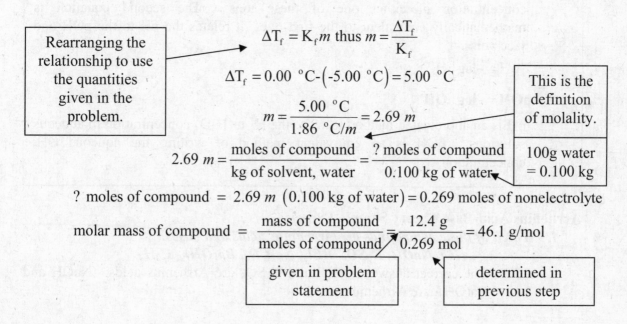

Rearranging the relationship to use the quantities given in the problem.

$$\Delta T_f = K_f m \text{ thus } m = \frac{\Delta T_f}{K_f}$$

$$\Delta T_f = 0.00 \ ^\circ C - \left(-5.00 \ ^\circ C\right) = 5.00 \ ^\circ C$$

This is the definition of molality.

$$m = \frac{5.00 \ ^\circ C}{1.86 \ ^\circ C/m} = 2.69 \ m$$

$$2.69 \ m = \frac{\text{moles of compound}}{\text{kg of solvent, water}} = \frac{? \text{ moles of compound}}{0.100 \text{ kg of water}}$$

100g water = 0.100 kg

? moles of compound = 2.69 m (0.100 kg of water) = 0.269 moles of nonelectrolyte

$$\text{molar mass of compound} = \frac{\text{mass of compound}}{\text{moles of compound}} = \frac{12.4 \text{ g}}{0.269 \text{ mol}} = 46.1 \text{ g/mol}$$

given in problem statement

determined in previous step

Module 17
Acids and Bases

Introduction

Two common theories of acids and bases commonly discussed in introductory chemistry texts are the Arrhenius and Brønsted-Lowry theories. There are also some important calculations necessary to determine the concentrations of species in aqueous acid, base, or buffer solutions. In this module the important concepts to understand are <u>a) the distinctions and commonalities between the two acid-base theories, b) how to distinguish between compounds that act as acids or bases, c) determining the hydronium and hydroxide concentrations in strong acid and base solutions, d) the pH, and pOH of some solutions, and e) what constitutes a buffer solution.</u>

Module 17 Key Equations & Concepts

1. **Acids produce H^+ in aqueous solution; bases produce OH^- in aqueous solution.**

   Arrhenius acid-base theory is the most restrictive theory requiring that the compound have an H^+ or OH^- and that it be dissolved in water.

2. **Acids are proton donors; bases are proton acceptors.**

   Brønsted-Lowry acid-base theory removes the restrictions that bases contain OH^- ions and that the solutions must be aqueous.

3. $$K_w = \left[H_3O^+ \right]\left[OH^- \right] = 1.00 \times 10^{-14}$$
   $$14 = pH + pOH$$

   K_w is the ionization constant for water. It is used to calculate either the hydronium or hydroxide ion concentration in aqueous solutions given the concentration of either one of these ions. The second equation is mathematically equivalent to the first one. It relates the pH to the pOH and vice versa.

4. $$pH = -\log \left[H^+ \right]$$
   $$pOH = -\log \left[OH^- \right]$$

   pH is an abbreviated method to write the H^+ or H_3O^+ concentration in aqueous solutions. pOH is an equivalent method of writing the aqueous OH^- concentration.

Arrhenius Acid-Base Theory Sample Exercise

1. ***Which of these compounds are Arrhenius acids and bases?***
   HCl, $NaOH$, H_2SO_4, BCl_3, Na_2CO_3, $Ba(OH)_2$, C_2H_4

   The correct answer is HCl and H_2SO_4 are Arrhenius acids. $NaOH$ and $Ba(OH)_2$ are Arrhenius bases.

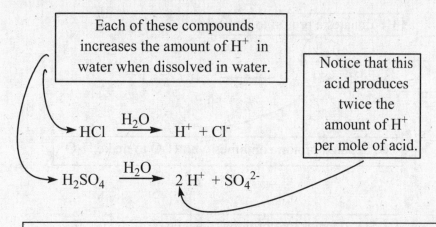

INSIGHT: To identify Arrhenius acids look for compounds that dissociate or ionize in water forming H^+.

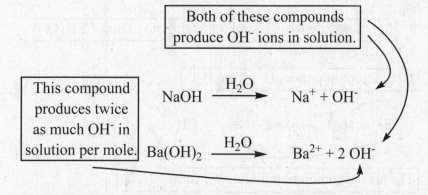

INSIGHT: To identify Arrhenius bases look for compounds that dissociate or ionize in water producing OH^-.

YIELD BF_3 and Na_2CO_3 are not Arrhenius acids or bases because there are no acidic H's or basic OH's. C_2H_4 has H's but they are not acidic because the C-H bond is too strong to be easily broken.

Brønsted-Lowry Acid-Base Theory Sample Exercise

2. *Which of these compounds can be classified as Brønsted-Lowry acids and bases?*

$$HF, NH_3, H_2SO_4, BCl_3, Na_2CO_3, K_2S$$

The correct answer is HF and H_2SO_4, are Brønsted-Lowry acids. NH_3, Na_2CO_3, and K_2S are Brønsted-Lowry bases.

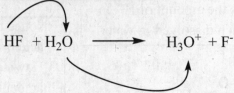

HF donates a proton to water.

$$HF + H_2O \longrightarrow H_3O^+ + F^-$$

The donated proton combines with H_2O to make H_3O^+.

H_2SO_4 donates two protons to water.

$$H_2SO_4 + 2 H_2O \longrightarrow 2 H_3O^+ + SO_4^{2-}$$

The donated protons combine with H_2O to make $2 H_3O^+$.

NH_3 accepts a proton, H^+, from H_2O.

$$NH_3 + H_2O \longrightarrow NH_4^+ + OH^-$$

The proton combines with NH_3 to form NH_4^+.

NH_3 accepts a proton, H^+, from HCl.

$$NH_3(g) + HCl(g) \longrightarrow NH_4Cl(s)$$

The proton combines with NH_3 to form NH_4^+. This reaction is an example of a <u>nonaqueous</u> Brønsted-Lowry acid-base reaction.

Carbonate ion, CO_3^{2-}, accepts a proton from water.

$$CO_3^{2-} + H_2O \longrightarrow HCO_3^- + OH^-$$

The proton combines with CO_3^{2-} to form HCO_3^-.
The Na^+ ions are spectator ions in this reaction.

YIELD | Because Arrhenius and Brønsted-Lowry acids are both proton donors, both are identified similarly. Brønsted-Lowry bases do not necessarily contain hydroxide ions, OH^-. It is possible for them to accept a proton from water to form hydroxide ions in aqueous solutions.

3. *Identify the Brønsted-Lowry acid-base conjugate pairs in these reactions.*

$$CH_3COOH + H_2O \rightleftharpoons CH_3COO^- + H_3O^+$$

$$F^- + H_2O \rightleftharpoons HF + OH^-$$

The correct answer is in the first reaction CH_3COOH acts as an acid and its conjugate base is CH_3COO^- while H_2O acts as a base with a conjugate acid of H_3O^+. In the second reaction, F^- acts as a base having the conjugate acid HF. In this reaction H_2O acts as an acid with the conjugate base OH^-.

CH_3COOH donates a proton to H_2O making it an acid. It's conjugate base, CH_3COO^-, differs from the acid by the loss of a single proton, H^+.

$$CH_3COOH + H_2O \rightleftharpoons CH_3COO^- + H_3O^+$$

H_2O accepts a proton from CH_3COOH making it a base. It's conjugate acid, H_3O^+, differs from the base by the addition of a single proton, H^+.

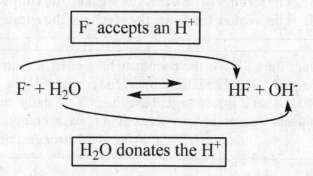

F^- accepts an H^+

$$F^- + H_2O \rightleftharpoons HF + OH^-$$

H_2O donates the H^+

$$CH_3COOH + H_2O \rightleftharpoons CH_3COO^- + H_3O^+$$

$$\text{acid}_1 \qquad \text{base}_2 \qquad\qquad \text{base}_1 \qquad \text{acid}_2$$

Notice that in one reaction H_2O is a base and in the other reaction it is an acid. **In Brønsted-Lowry theory all acid-base reactions are a competition between stronger and weaker acids or bases.** In the CH_3COOH reaction, the stronger acid is CH_3COOH, thus water acts as a base in the presence of CH_3COOH. In the F^- reaction, H_2O is the stronger acid so it acts as the acid in this reaction. Water is an amphoteric species; it acts as either an acid or a base in the presence of a stronger acid or base.

4. *Arrange the following species in order of their strength as bases.*
   HCO_3^-, Cl^-, CO_3^{2-}
   The correct answer is $Cl^- < HCO_3^- < CO_3^{2-}$.

INSIGHT:

The best approach to a problem like this is to recognize which acid or base the species are derived from. The Cl^- ion is derived from the acid HCl, HCO_3^- derives from H_2CO_3, and CO_3^{2-} derives from HCO_3^-. (In each case the acid the species derives from is determined by adding H^+ to the species. In other words, determine the conjugate acids of each species.) Now we can easily compare the strengths of the conjugate acids and determine the base strengths. For acid strengths, HCl is by far the strongest acid, H_2CO_3 is the next strongest acid, and finally the HCO_3^- ion is the weakest acid. In fact, the HCO_3^- ion is a weak base.

Some important things to remember about acids and bases are:
   1) **The stronger the acid, the weaker the conjugate base.**
   2) **The weaker the acid, the stronger the conjugate base.**
   3) **The stronger the base, the weaker the conjugate acid.**
   4) **The weaker the base, the stronger the conjugate acid.**

INSIGHT:

Since the Cl^- ion is the conjugate base of HCl, a strong acid, it is the weakest base. The HCO_3^- ion is the conjugate base of the weak acid H_2CO_3, thus it is a stronger base than Cl^-. Finally, the CO_3^{2-} ion is the conjugate base of the very weak acid (a basic compound is a very weak acid) HCO_3^- making it the strongest base.

118

Water Ionization Constant Sample Exercises

5. ***What is the [OH⁻] in an aqueous solution that has a pH of 5.25?***

The correct answer is $[OH^-] = 1.8 \times 10^{-9}$ M.

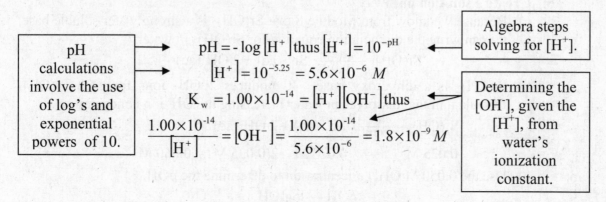

pH calculations involve the use of log's and exponential powers of 10.	Algebra steps solving for $[H^+]$.

$$pH = -\log[H^+] \text{ thus } [H^+] = 10^{-pH}$$

$$[H^+] = 10^{-5.25} = 5.6 \times 10^{-6} \ M$$

$$K_w = 1.00 \times 10^{-14} = [H^+][OH^-] \text{ thus}$$

$$\frac{1.00 \times 10^{-14}}{[H^+]} = [OH^-] = \frac{1.00 \times 10^{-14}}{5.6 \times 10^{-6}} = 1.8 \times 10^{-9} \ M$$

Determining the $[OH^-]$, given the $[H^+]$, from water's ionization constant.

6. ***What is the pH of an aqueous solution that has a [OH⁻] = 3.45 x 10⁻³?***

The correct answer is pH = 11.538.

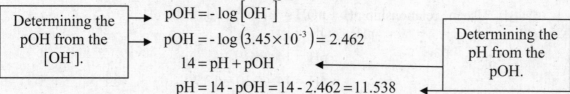

Determining the pOH from the $[OH^-]$.

$$pOH = -\log[OH^-]$$

$$pOH = -\log(3.45 \times 10^{-3}) = 2.462$$

$$14 = pH + pOH$$

$$pH = 14 - pOH = 14 - 2.462 = 11.538$$

Determining the pH from the pOH.

Strong Acid/Strong Base Dissociation Sample Exercises

7. ***What is the pH of an aqueous 0.025 M HCl solution?***

The correct answer is pH = 1.60.

Because HCl is a strong acid, it ionizes 100% in water. All 0.025 moles of the HCl molecules ionize in water to produce H⁺ and Cl⁻ ions as shown in the following equation.

$$HCl(g) \xrightarrow[H_2O]{100\%} H^+(aq) + Cl^-(aq)$$

0.025 M   0.025 M   0.025M

Thus the $[H^+] = 0.025$ M and pH = -log $[H^+]$ = 1.60.

8. ***What is the pH of an aqueous 0.025 M Sr(OH)₂ solution?***

The correct answer is pH = 12.70.

This exercise is slightly more complex than exercise 7. Our problem solving steps check list will help explain the major differences.

1. **Read the problem**
2. **Understand precisely what the problem is asking**
   a) We are to calculate the pH of an aqueous $Sr(OH)_2$ solution.
3. **Identify the chemical principles**
   a) $Sr(OH)_2$ is a strong base.
   b) Understand that $Sr(OH)_2$ dissociates into 1 Sr^{2+} ion and 2 OH^- ions.
   c) Proper use of the pH and pOH relationships.
4. **Determine the relevant pieces of information**

a) The most important pieces of information are
1) Concentration of 0.025 M
2) Formula of $Sr(OH)_2$

5. **Create a solution pathway**
    a) Because we know from Module 8 that $Sr(OH)_2$ is a strong water soluble base, we can write the dissociation equation for $Sr(OH)_2$ in water.

$$Sr(OH)_2 \xrightarrow[H_2O]{100\%} Sr^{2+}(aq) + 2\,OH^-(aq)$$

    b) $Sr(OH)_2$ is a dihydroxy base. It produces 2 OH- ions for each $Sr(OH)_2$ formula unit that dissolves in water increasing the OH^- ion concentration.

$$Sr(OH)_2 \xrightarrow[H_2O]{100\%} Sr^{2+}(aq) + 2\,OH^-(aq)$$

    0.025 M   0.025 M   2(0.025 M) = 0.050 M

    c) Use the 0.050 M OH^- concentration to determine the pOH.

$$pOH = -\log[OH^-]$$
$$pOH = -\log(0.050)$$
$$pOH = 1.30$$

    d) Use the relationship pH + pOH = 14 to calculate the pH.
$$pH + pOH = 14$$
$$pH = 14 - pOH$$
$$pH = 14 - 1.30 = 12.70$$

6. **Check your answer**
    a) $Sr(OH)_2$ forms a strongly basic solution.
    b) A pH of 12.70 is consistent with a strongly basic solution.
    c) Our answer is correct.

INSIGHT:
> Strontium hydroxide, $Sr(OH)_2$, is a water soluble, strong, polyhydroxy base. When it dissociates in water, the $[OH^-]$ in solution is twice the molarity of the $Sr(OH)_2$. This is also true for $Ca(OH)_2$ and $Ba(OH)_2$. There is one strong polyprotic acid to be aware of, H_2SO_4. For H_2SO_4, the $[H^+]$ will be nearly twice the molarity.

YIELD
> In aqueous acid-base problems, you must first decide if the solution is composed of a strong acid or strong base. Strong acid and base solutions are easy to calculate because the acids and bases ionize 100% in water. Watch for polyprotic acids and polyhydroxy bases as they have increased ion concentrations.

Buffer Solution Sample Exercise

9. *Which of these aqueous solutions are buffers?*
    *   *a. 0.035 M HCl dissolved in 0.025 M NaCl*
    *   *b. 0.035 M CH_3COOH dissolved in 0.025 M NaCl*
    *   *3) 0.035 M CH_3COOH dissolved in 0.025 M $NaCH_3COO$*
    *   *4) 0.035 M KOH dissolved in 0.025 M KCl*
    *   *5) 0.035 M KOH dissolved in 0.025 M NH_4Cl*
    *   *6) 0.035 M NH_3 dissolved in 0.025 M NH_4Cl*

    The correct answer is solutions c. and f. are buffers.

INSIGHT:	Buffer solutions are made by mixing a weak acid with the salt of its conjugate base or a weak base with the salt of its conjugate acid. In solution c. weak acetic acid, CH_3COOH, is mixed with the sodium salt of its conjugate base, $NaCH_3COO$. In solution f. the weak base NH_3 is mixed with the chloride salt of its conjugate acid, NH_4Cl.

Module 18
Chemical Equilibrium

Introduction
In this module we will examine some of the basic ideas and calculations required for chemical equilibrium problems. In this module we will examine a) collision theory and how it helps us to understand the effects of catalysis, b) the purpose and uses of the equilibrium constant (K_c), c) how to use Le Châtelier's Principle to predict how equilibrium reactions respond to stresses, d) how K_c is determined for heterogeneous reactions, and e) how to use K_{sp} in solid-aqueous solution equilibria.

Module 18 Key Equations & Concepts

1. **E_a – Activation Energy**
   Energy required to initiate reaction between molecules.

2. $K_c = \dfrac{[C]^c [D]^d}{[A]^a [B]^b}$ **for the reaction** $aA + bB \rightleftarrows cC + dD$

   The equilibrium constant, K_c, is used to determine whether the reaction is product or reactant favored (i.e. yields a preponderance of reactants or products) and to determine the concentrations of the reactants and products at equilibrium. Concentrations used in K_c must be equilibrium concentrations. The symbols a, b, c, and d are the stoichiometric coefficients for the reaction.

3. $K_{sp} = [M^+]^x [Y^-]^z$ **for the dissolution** $M_x Y_z \rightarrow x\,M^+ + z\,Y^-$

   Equilibrium constant for solids that are slightly water soluble.

Collision Theory Sample Exercise

1. *Choose the statements which, according to collision theory, are appropriate conditions for a chemical reaction to occur.*
   a. *Reactant molecules must have translational motion.*
   b. *Reactant molecules must collide.*
   c. *Reactant molecules must have sufficient energy to initiate reaction.*
   d. *Reactant molecules must be properly oriented.*
   e. *Reactant molecules must be in gas phase.*
      The correct answer is responses b, c, and d.

INSIGHT:

Collision theory describes how molecules must interact to ensure that a chemical reaction occurs. There are three necessary conditions for molecules to react: 1) the molecules must collide with one another, 2) the molecules must collide with sufficient energy to initiate reaction, and 3) the molecules must be oriented so that the breaking and reforming of chemical bonds can occur. The energy necessary to initiate a reaction is called the energy of activation, E_a. It is the subject of exercise 2.

Energy of Activation and Catalysis Sample Exercises

2. *Which numbered arrow in this potential energy diagram indicates the activation energy for the forward reaction?*

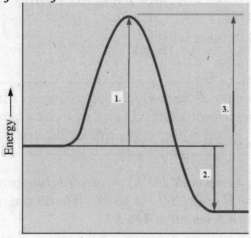

The correct answer is arrow 1.

3. *The following two potential energy diagrams are for the same chemical reaction. One of these diagrams is for a chemical reaction that has been catalyzed and the other is for a reaction that has not been catalyzed. Which diagram depicts the catalyzed reaction?*

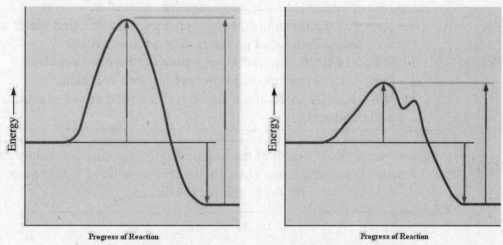

The correct answer is the image on the right is the catalyzed reaction.

INSIGHT: E_a, the activation energy, is the "hill of energy" that molecules must climb to initiate reaction. In potential energy diagrams, it is the large hill between the reactants and products. Frequently, catalysts decrease the activation energy, which speeds up the reaction. In the two diagrams above, the one on the right has a significantly decreased activation energy and, therefore, representative of a catalyzed reaction's potential energy diagram.

Use of the Equilibrium Constant, K$_c$, Sample Exercises

4. *What is the form of the equilibrium constant, K$_c$, for this reaction?*

$$H_2(g) + I_2(g) \rightleftharpoons 2\ HI(g)$$

The correct answer is $\dfrac{[HI]^2}{[H_2][I_2]}$.

INSIGHT:	K$_c$ is the ratio of the product concentrations raised to the power of their stoichiometric coefficients divided by the reactant concentrations raised to the power of their stoichiometric coefficients.

5. *For the following reaction at 298 K, the equilibrium concentrations are [H$_2$] = 1.50 M, [I$_2$] = 2.00 M, and [HI] = 3.46 M. What is the value of the equilibrium constant, K$_c$, for this reaction at 298 K?*

$$H_2(g) + I_2(g) \rightleftharpoons 2\ HI(g)$$

The correct answer is 4.00.

Units are not used in equilibrium constants. We are interested in K$_c$'s size.	For this reaction $K_c = \dfrac{[HI]^2}{[H_2][I_2]}$ thus $K_c = \dfrac{[3.46]^2}{[1.50][2.00]} = \dfrac{12.0}{3.00} = 4.00$	A very common mistake is to forget to properly include the stoichiometric coefficients as exponents.

YIELD	The size of an equilibrium constant indicates if the reaction yields a preponderance of products, reactants, or neither. 1) If **K$_c$ >10 to 20**, the reaction is a **product favored reaction**. 2) If **K$_c$ < 1**, the reaction is a **reactant favored reaction**. 3) If **1 < K$_c$ < 10 to 20**, the reaction yields a **mixture of reactants and products**.

INSIGHT:	In one sense, K$_c$ is a ratio of the product concentrations divided by the reactant concentrations. Thus, the larger the value of K$_c$ the more products relative to reactants.

6. *Shown below are molecular representations of four mixtures of H₂, I₂, and HI. At 298 K the Kc for this reaction H₂(g) + I₂(g) ⇌ 2 HI(g) has a value of 4.00. Which representations show equilibrium mixtures?*

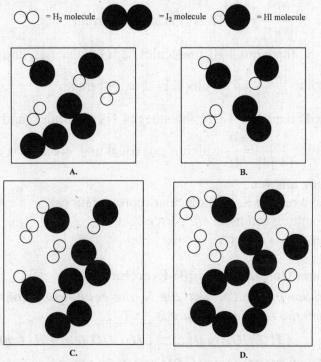

The correct answer is pictures B and C are equilibrium mixtures.

Let's examine this exercise using the problem solving steps check list.

1. **Read the problem**
2. **Understand precisely what the problem is asking**
   a) This problem is asking us to relate the numerical understanding of K_c to a molecular representation of the idea.
3. **Identify the chemical principles**
   a) Understanding equilibrium.
   b) Understand what K_c is telling us about the ratio of products to reactants.
   c) Relating K_c to a molecular representation of equilibrium.
4. **Determine the relevant pieces of information**
   a) The one irrelevant piece of information is the temperature. We shall not use it in our solution. However, K_c would have a different value at another temperature.
5. **Create a solution pathway**

   a) The K_c for this reaction is $\dfrac{[HI]^2}{[H_2][I_2]}$. Thus we must count the number of H₂, I₂, and HI molecules in each representation then place those values into $\dfrac{[HI]^2}{[H_2][I_2]}$ and see if the answer is 4.

125

b) In box A there are 2 HI molecules, 2 H_2 molecules, and 2 I_2 molecules. This results in $\dfrac{[2]^2}{[2][2]} = 1$ which does not equal 4. Thus A is not at equilibrium.

c) In box B there are 2 HI molecules, 1 H_2 molecules, and 1 I_2 molecules. This results in $\dfrac{[2]^2}{[1][1]} = 4$. Thus B is at equilibrium.

d) In box C there are 4 HI molecules, 2 H_2 molecules, and 2 I_2 molecules. This results in $\dfrac{[4]^2}{[2][2]} = 4$. Thus C is at equilibrium.

e) In box D there are 3 HI molecules, 4 H_2 molecules, and 4 I_2 molecules. This results in $\dfrac{[3]^2}{[4][4]} = \dfrac{9}{16}$ which is not equal to 4. Thus D is not at equilibrium.

6. **Check your answer**
   a) We have related K_c to molecular representations.
   b) Our counting and math are correct.
   c) Our answer is correct.

Size of Equilibrium Constants Sample Exercise

7. *For the following reaction at 298 K, the equilibrium constant is 1.8×10^{-5}. At equilibrium the reaction mixture is _____ ?*

$$CH_3COOH(aq) \rightleftharpoons CH_3COO^-(aq) + H^+(aq)$$

   a. *100% CH_3COOH and 0% CH_3COO^- and H^+*

   b. *0% CH_3COOH and 100% CH_3COO^- and H^+*

   c. *50% CH_3COOH and 50% CH_3COO^- and H^+*

   d. *mostly CH_3COOH and a little CH_3COO^- and H^+*

   e. *a little CH_3COOH and mostly CH_3COO^- and H^+*

   The correct answer is d mostly CH_3COOH and a little CH_3COO^- and H^+.

INSIGHT:	The small value of the equilibrium constant, 1.8×10^{-5}, indicates that this equilibrium is reactant favored giving few products, CH_3COO^- and H^+, and significant amounts of reactants, CH_3COOH.

Heterogeneous Equilibrium Sample Exercise

8. **Choose the correct form of the equilibrium constant for this reaction.**

$$4 Fe(s) + 3 O_2(g) \rightleftharpoons 2 Fe_2O_3(s)$$

a. $K = \dfrac{[Fe_2O_3]^4}{[Fe]^4 [O_2]^3}$

b. $K = \dfrac{[Fe_2O_3]}{[Fe] [O_2]}$

c. $K = \dfrac{1}{[Fe]^4 [O_2]^3}$

d. $K = \dfrac{[Fe_2O_3]^4}{[Fe]^4}$

e. $K = \dfrac{1}{[O_2]^3}$

The correct answer is e. $K = \dfrac{1}{[O_2]^3}$.

INSIGHT: K_c is actually defined using a thermodynamic quantity called activity. The activities of gases are the same as their concentrations. In heterogeneous equilibria (those involving gases, liquids, and solids) the activities of the pure solids and liquids are 1 and can be neglected. Thus the K_c for heterogeneous equilibria will only require the gas's concentrations and not any solids or liquids involved in the equilibrium.

Le Châtelier's Principle Sample Exercise

9. **What would be the effect of each of these changes on the position of equilibrium of this reaction at 298 K?**

$$2 NO_2(g) \rightleftharpoons N_2O_4(g) \quad \Delta H^0_{rxn} = -57.2 \text{ kJ/mol}$$

a. **Increasing the temperature of the reaction**
b. **Removing some NO_2 from the reaction vessel.**
c. **Adding some N_2O_4 to the reaction vessel.**
d. **Increasing the pressure in the reaction vessel by adding an inert gas.**
e. **Decreasing by half the size of the reaction vessel.**
f. **Introducing a catalyst into the reaction vessel.**

The correct answers are the position of equilibrium will shift to the a) left, b) left, c) left, d) no effect, e) right, and f) no effect.

INSIGHT: If the position of equilibrium shifts to the left the reactant concentrations increase and the product concentrations decrease. If the position of equilibrium shifts to the right the reactant concentrations decrease and the product concentrations increase. All of the following changes are illustrations of Le Châtelier's principle:

---

a) ***For exothermic reactions, increasing the temperature shifts the position of equilibrium to the left, decreasing the temperature shifts the position of equilibrium the right. Endothermic reactions behave oppositely.*** In exercise 8, the negative ΔH^0_{rxn} tells us that the reaction is exothermic, thus increasing the temperature shifts the position of equilibrium to the left.

b) ***If a reactant's concentration is decreased below the equilibrium concentration, the position of equilibrium will change to restore concentrations that correspond to those predicted by the equilibrium constant.*** In this exercise, removing some NO_2 from the reaction vessel decreases the $[NO_2]$. The reaction equilibrium responds to this stress by increasing the $[NO_2]$ and decreasing the $[N_2O_4]$, an equilibrium position shift to the left or reactant side. Adding NO_2 would cause the position of equilibrium to shift to the right or product side.

c) ***If a product's concentration is increased above the equilibrium concentration, the equilibrium position will shift to restore concentrations of products and reactants that correspond to those predicted by the equilibrium constant.*** Adding some N_2O_4 to the reaction vessel increases the $[N_2O_4]$ above the equilibrium concentration. The reaction equilibrium responds to this stress by decreasing the $[N_2O_4]$ and increasing the $[NO_2]$, an equilibrium position shift to the left or reactant side. Removing N_2O_4 would shift the position of equilibrium to the right or product side.

d) ***Adding an inert gas to the reaction mixture has no effect on the equilibrium position because the concentrations of the gases are not changed.*** This is a common misconception for students.

e) ***If the volume of the reaction vessel is changed, the concentrations of the gases are changed because for gases, M ∝ n/V. If the vessel's volume is decreased, the equilibrium position will shift to the side that has the fewest moles of gas.*** In this exercise the right or product side has the fewest moles.

f) ***Adding a catalyst has no effect on the position of equilibrium.*** Catalysts change the rates of reactions but not positions of equilibrium.

128

Solubility Product Sample Exercises

10. The solubility of iron(II) hydroxide, Fe(OH)₂, in water is 1.1×10^{-3} g/L at 25.0°C. What is the solubility product constant for Fe(OH)₂?

The correct answer is $K_{sp} = 6.9 \times 10^{-15}$.

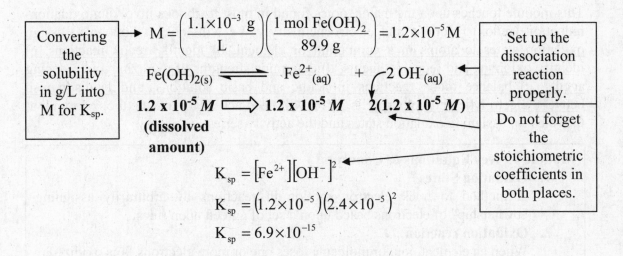

11. What are the molar solubilities of Zn²⁺ and OH⁻ for zinc hydroxide at 25.0°C? $K_{sp} = 4.5 \times 10^{-17}$

The correct answer is $[Zn^{2+}] = 2.2 \times 10^{-5}$ M, $[OH^-] = 4.4 \times 10^{-5}$ M.

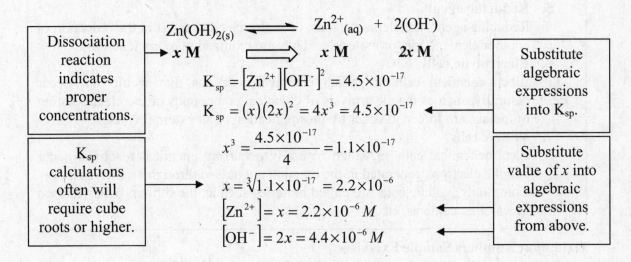

Module 19
Oxidation-Reduction Reactions and Electrochemistry

Introduction

This module touches upon the major topics found in most textbooks involving oxidation-reduction (redox) reactions and electrochemistry. We will look at how to <u>a) assign oxidation states to atoms in a compound or element, b) identify redox reactions, c) identify oxidizing and reducing agents, d) determine strengths of oxidizing and reducing agents, e) balance redox reactions in acidic and basic solutions, and f) distinguish between electrolytic and voltaic cells.</u> You will need access to your textbook to look at the rules for assigning oxidation states and the activity series.

Module 19 Key Equations & Concepts

1. **Oxidation States**
   A method to track electron motion in reactions by arbitrarily assigning "ownership" of electrons based upon a set of agreed upon rules.

2. **Oxidation reaction**
   When an element, ion, or molecule loses one or more electrons, it is oxidized.

3. **Reduction reaction**
   When an element, ion, or molecule gains one or more electrons, it is reduced.

4. **Oxidizing agent**
   Oxidizing agents are elements, ions, or molecules that assist in the oxidation of another element, ion, or molecule. They are reduced in redox reactions.

5. **Reducing agent**
   Reducing agents are elements, ions, or molecules that assist in the reduction of another element, ion, or molecule. They are oxidized in redox reactions.

6. **Electrolytic cells**
   Electrochemical cells in which chemical reactions that would not occur naturally, such as the electrolysis of chemical compounds or the electroplating of metals, are forced to occur by the application of an external voltage.

7. **Voltaic cells**
   Electrochemical cells in which naturally occurring chemical reactions occur and the electrons generated in the reaction are passed through an external wire. Commonly, voltaic cells are called batteries such as the lithium batteries used in watches, cameras, etc.

Oxidation Numbers Sample Exercises

1. ***What are the oxidation numbers of each element in $HClO_4$?***
   > The correct answer is H is in the 1+ oxidation state, O is in the 2- oxidation state, and Cl is in the 7+ oxidation state.

> In your textbook there is a set of rules indicating how to assign oxidation numbers to elements in a molecule or ion. You must know those rules by heart to perform these techniques.

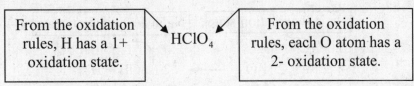

From the oxidation rules, H has a 1+ oxidation state.

$HClO_4$

From the oxidation rules, each O atom has a 2- oxidation state.

Since we know the oxidation states of H and O from the rules, we use those values to calculate the Cl oxidation state. **In a neutral compound, the sum of the oxidation states of each atom must equal zero.** A small amount of algebra gives us the oxidation state of Cl.

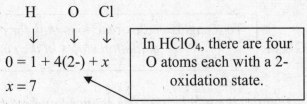

$$H \quad O \quad Cl$$
$$\downarrow \quad \downarrow \quad \downarrow$$
$$0 = 1 + 4(2\text{-}) + x$$
$$x = 7$$

In $HClO_4$, there are four O atoms each with a 2- oxidation state.

INSIGHT: Chemists refer to oxidation states on a per atom basis. While there are four O atoms in $HClO_4$ each with a 2- oxidation state, we do not say that O has an 8- oxidation state.

INSIGHT: Notice that Cl in $HClO_4$ does **not** have an oxidation state of 1-. Oxidation states are not the same as an ionic charge. In most of its ionic compounds, Cl by itself has a -1 ionic charge. However, in combination with other elements Cl can have oxidation states ranging from 7- to 7+.

2. *What are the oxidation numbers of each element in the sulfite ion, SO_3^{2-}?*

The correct answer is O is in the 2- oxidation state, and S is in the 4+ oxidation state.

Immediately, we know the oxidation state of O is 2- from the rules. **In an ion, the sum of the oxidation states of each atom must equal the ion's charge.** We use those two values to calculate the S oxidation state.

$$O \quad S$$
$$\downarrow \quad \downarrow$$
$$-2 = 3(2\text{-}) + x$$
$$x = 4$$

Reduction Oxidation Reaction Sample Exercises

3. *What reaction types are represented by this chemical reaction?*
$$2\,Na \;+\; 2\,H_2O \;\rightarrow\; 2\,NaOH \;+\; H_2$$
The correct answer is reduction oxidation (redox) and combination reaction.

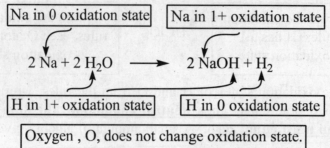

4. *Identify the species that are reduced and oxidized in this chemical reaction.*

$$2\,Na \;+\; 2\,H_2O \;\rightarrow\; 2\,NaOH \;+\; H_2$$

The correct answer is H is reduced and Na is oxidized.

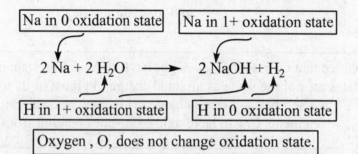

5. *Identify the oxidizing and reducing agents in this chemical reaction.*

$$2\,Na \;+\; 2\,H_2O \;\rightarrow\; 2\,NaOH \;+\; H_2$$

The correct answer is H is the oxidizing agent and Na is the reducing agent.

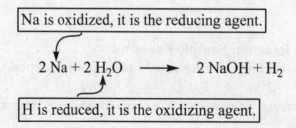

Oxidation and reduction cannot occur in isolation. For one species to be oxidized, another species must be reduced. In this reaction, Na, which is oxidized, is the agent that helps to reduce H atoms in water molecules. That is why Na is designated as the reducing agent. Likewise, the reduction of H assists the oxidation of Na, so H is the oxidizing agent. **In all cases, the species that is oxidized is the reducing agent and the reduced species is the oxidizing agent.**

Strength of Oxidizing and Reducing Agents Sample Exercise

6. *Arrange the following species in order of increasing reducing agent strength.*
   Li, Au, Cu, Zn
   The correct answer is Au < Cu < Zn < Li.

Shown below is a portion of the activity series similar to the one in your textbook.

Activity Series
Li
Ba
Mg
Zn
Fe
Pb
Cu
Ag
Au

INSIGHT:

The activity series is a tabulation of substances in order of oxidation ability. In the abbreviated table shown above, Li is the most easily oxidized substance and Au is the least easily oxidized (most easily reduced) substance. A substance that is easily oxidized is a strong reducing agent while one that is easily reduced is a very weak reducing agent. Thus the four metals in exercise 6 in order of their ability to be oxidized, which is the same as their order of reducing agent strength, are
Au < Cu < Zn < Li.

Balancing Redox Reactions Sample Exercises

7. *Balance the following redox reaction in acidic solution.*
   $$Cu(s) + NO_3^-(aq) \rightarrow Cu^{2+}(aq) + NO_2(g)$$
   The correct answer is $Cu + 4 H^+ + 2 NO_3^- \rightarrow Cu^{2+} + 2 NO_2 + 2 H_2O$.

INSIGHT:

Redox reactions occur in either acidic or basic solutions. In acidic solutions, you can add H^+ and H_2O to balance the reaction. In basic solutions, you can add OH^- and H_2O. The problem will either state that the reaction is in acidic solution or H^+ will be present as a reactant or product. Similarly, for basic solutions look for a statement that the reaction is in basic solution or the presence of OH^-.

There are two simple methods to balance redox reactions, the change in oxidation number method and the half-reaction method. The change in oxidation number method is more physically correct but the half-reaction method is simpler to learn and more straight forward. We will use the half-reaction method in these exercises. It is very important that you review the rules for assigning oxidation numbers and for balancing redox reactions found in your textbook.

These are multistep problems which are well suited for our problem solving steps check list.

1. **Read the problem**
2. **Understand precisely what the problem is asking**
   a) We must balance this reaction $Cu(s) + NO_3^-(aq) \rightarrow Cu^{2+}(aq) + NO_2(g)$ in acidic solution.
3. **Identify the chemical principles**
   a) Balancing redox reactions
4. **Determine the relevant pieces of information**
   a) There are two important pieces of information
      1) The reaction $Cu(s) + NO_3^-(aq) \rightarrow Cu^{2+}(aq) + NO_2(g)$
      2) The solution is acidic
5. **Create a solution pathway**
   a) Separate the reaction into two half-reactions.

$$Cu \rightarrow Cu^{2+}$$

$$NO_3^- \rightarrow NO_2$$

   b) Balance each half-reaction by mass. Because the solution is acidic, we can add H^+ or H_2O as necessary.

$$Cu \rightarrow Cu^{2+}$$

$$2\,H^+ + NO_3^- \rightarrow NO_2 + H_2O$$

   c) Balance each half-reaction by charge. Add electrons to each half-reaction to ensure that the charge on each side of the half-reactions is equal.

$$Cu \rightarrow Cu^{2+} + 2e^-$$

$$e^- + 2\,H^+ + NO_3^- \rightarrow NO_2 + H_2O$$

   d) Notice that the Cu half-reaction has 2 e^- as products while the NO_3^- half-reaction has 1 e^- as a reactant. In the final balanced reaction, the number of reactant electrons must equal the number of product electrons. To accomplish this, double the NO_3^- half-reaction.

$$Cu \rightarrow Cu^{2+} + 2e^-$$

$$2\,[e^- + 2\,H^+ + NO_3^- \rightarrow NO_2 + H_2O]$$

134

e) Add the two half-reactions. Since the reactant electrons equal the product electrons, they cancel out in the final reaction.

$$Cu \rightarrow Cu^{2+} + 2e^-$$

$$2 \left[e^- + 2 H^+ + NO_3^- \rightarrow NO_2 + H_2O \right]$$

$$Cu + 4 H^+ + 2 NO_3^- \rightarrow Cu^{2+} + 2 NO_2 + 2 H_2O$$

6. Check your answer
   a) There are 2 Cu atoms, 2 N atoms, 4 H atoms, and 6 O atoms on both sides of the reaction. The reaction is mass balanced.
   b) The total charge on the left side is 2+ as is the case on the right side. The reaction is charge balanced.
   c) Our answer is correct.

8. *Balance the following redox reaction in basic solution.*

$$CrO_2^-(aq) + ClO^-(aq) \rightarrow CrO_4^-(aq) + Cl^-(aq)$$

The correct answer is $CrO_2^- + 2ClO^- \rightarrow CrO_4^- + 2Cl^-$.

1) Separate the reaction into the reduction and oxidation half reactions.

$$CrO_2^- + ClO^- \rightarrow CrO_4^- + Cl^-$$
$$a)\ CrO_2^- \rightarrow CrO_4^-$$
$$b)\ CrO_2^- + 4OH^- \rightarrow CrO_4^- + 2H_2O$$
$$c)\ CrO_2^- + 4OH^- \rightarrow CrO_4^- + 2H_2O + 4e^-$$
$$d)\ ClO^- \rightarrow Cl^-$$
$$e)\ ClO^- + H_2O \rightarrow Cl^- + 2OH^-$$
$$f)\ ClO^- + H_2O + 2e^- \rightarrow Cl^- + 2OH^-$$
$$g)\ CrO_2^- + 4OH^- \rightarrow CrO_4^- + 2H_2O$$
$$h)\ 2ClO^- + 2H_2O \rightarrow 2Cl^- + 4OH^-$$
$$i)\ CrO_2^- + 4OH^- + 2ClO^- + 2H_2O \rightarrow CrO_4^- + 2H_2O + 2Cl^- + 4OH^-$$
$$j)\ CrO_2^- + 2ClO^- \rightarrow CrO_4^- + 2Cl^-$$

2) Balance H and O atoms in half-reaction a) by adding OH⁻ and H₂O.
3) Add 4 e⁻ to step b) to balance charge.
5) Add 2 e⁻ to step e) to balance charge.
6) Step c) plus twice step f).
8) Remove 4 OH⁻ and 2 H₂O from both sides.

4) Balance H and O atoms in half-reaction d) by adding OH⁻ and H₂O.
7) Add steps g) and h).

Electrolytic and Voltaic Cell Sample Exercises
   9. *The battery in a watch is an example of a(n) _____ cell. Silverware is made by chemically placing a thin layer of silver on a base metal using a(n) _____ cell.*
      The correct answer is voltaic and electrolytic.

135

10. *A voltaic cell is constructed of a 1.0 M $CuSO_4$(aq) solution and a 1.0 M $AgNO_3$(aq) solution plus the electrodes, connecting wires, and salt bridges. What chemical species will be produced at the cathode and anode? What is the direction of electron flow in this cell?*

The correct answer is Cu is oxidized to Cu^{2+} at the anode and Ag^+ is reduced to Ag at the cathode. The electrons flow from the anode to the cathode in voltaic cells. Because the reactions occur naturally, no external source of electricity is needed.

In all voltaic cells the electrons flow from the anode (negative electrode) to the cathode (positive electrode).

The salt bridge allows the counter ions to pass between the cells and keep the solutions neutral.

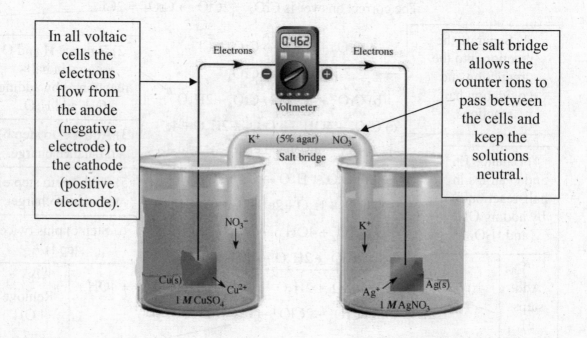

Module 20
Nuclear Chemistry

Introduction
Module 20 describes some of the basic relationships discussed in typical nuclear chemistry chapters. The important topics described are a) how to predict the products of alpha, negatron, and positron radioactive decays and the products of nuclear reactions, b) simple half-life problems, c) some concepts of fission and fusion, and d) a brief discussion of radiation damage to human tissue.

Module 20 Key Equations & Concepts

1. $^{A}_{Z}X$

   Isotopic symbols are written using this format. X is the elemental symbol. Z is the proton number. A is the atomic mass number which equals the proton number plus the neutron number, Z + N.

2. $^{A}_{Z}X \rightarrow {}^{A-4}_{Z-2}Y + {}^{4}_{2}He$

   Radioactive alpha decay removes two protons and two neutrons, in the form of a ^{4}He nucleus, from the decaying nucleus and converts the element X into a different element Y.

3. $^{A}_{Z}X \rightarrow {}^{A}_{Z+1}Y + {}^{0}_{-1}e$ (or $^{0}_{-1}\beta^{-}$)

   Radioactive beta decay, β^{-} or negatron decay, converts a neutron into a proton by eliminating a high velocity electron, the β^{-} particle, from the nucleus. The decaying nucleus, X, is converted to a new nucleus, Y, having one additional proton and one less neutron.

4. $^{A}_{Z}X \rightarrow {}^{A}_{Z-1}Y + {}^{0}_{+1}e$ (or $^{0}_{+1}\beta^{+}$)

   Radioactive positron decay, β^{+}, converts a proton into a neutron by eliminating a high velocity positive electron, the β^{+} particle, from the nucleus. The decaying nucleus, X is converted to a different element, Y, having one less proton and one more neutron.

5. $^{M_1}_{Z_1}Q \rightarrow {}^{M_2}_{Z_2}R + {}^{M_3}_{Z_3}Y$ where $M_1 = M_2 + M_3$ and $Z_1 = Z_2 + Z_3$

   This is the basic relationship for nuclear reactions and radioactive decay. The proton numbers of the product nuclides (Z_2 and Z_3) must sum up to the original nuclide's proton number, Z_1. The mass numbers of the product nuclides (M_2 and M_3) must also add up to the original nuclide's mass, M_1.

6. $t_{1/2}$ or half-life

   Half-life is the time necessary for ½ of the starting material to radioactively decay. A 100.0 g sample of a radioactive substance with a $t_{1/2}$ of 0.50 hours has 50.0 g of the sample remaining after 30 minutes.

Alpha Decay Sample Exercise
1. **What is the product nuclide of the alpha decay of ^{232}Th?**

   The correct answer is ^{228}Ra.

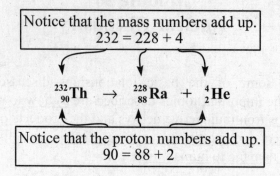

Notice that the mass numbers add up.
232 = 228 + 4

$$^{232}_{90}\text{Th} \rightarrow {}^{228}_{88}\text{Ra} + {}^{4}_{2}\text{He}$$

Notice that the proton numbers add up.
90 = 88 + 2

INSIGHT: Alpha decay occurs primarily in nuclides that have more than 83 protons. To determine the product nuclide, take the proton number of the decaying nucleus and subtract 2. The product nuclide's mass number is the decaying nuclide's mass number minus 4.

Beta Decay Sample Exercises

2. ***What is the product nuclide of the β^-, negatron, decay of ^{14}C?***
The correct answer is ^{14}N.

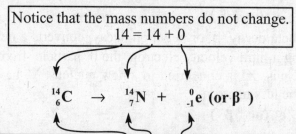

Notice that the mass numbers do not change.
14 = 14 + 0

$$^{14}_{6}\text{C} \rightarrow {}^{14}_{7}\text{N} + {}^{0}_{-1}\text{e (or } \beta^-)$$

Notice that the charges of the protons and the beta particle add up.
6 = 7 + (-1)

3. ***What is the product nuclide of the β^+, positron, decay of ^{37}Ca?***
The correct answer is ^{37}K.

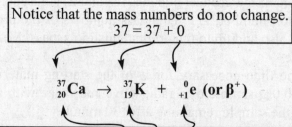

Notice that the mass numbers do not change.
37 = 37 + 0

$$^{37}_{20}\text{Ca} \rightarrow {}^{37}_{19}\text{K} + {}^{0}_{+1}\text{e (or } \beta^+)$$

Notice that the charges of the protons and the beta particle add up.
20 = 19 + (+1)

INSIGHT:	1) In all forms of beta decay, the mass numbers do not change.
	2) In β^- decay, the product nuclide has one **more** proton than the decaying nuclide.
	3) In β^+ decay, the product nuclide has one **less** proton than the decaying nuclide.

Nuclear Reaction Sample Exercise

4. *Fill in the missing nuclide in this nuclear reaction.*

$$^{53}Cr + {}^{4}He \rightarrow \underline{\quad} + 2\,n$$

The correct answer is ^{55}Fe.

> The mass number is determined from the sum of the mass numbers of the reactants and products.
> $53 + 4 = x + 2$ thus $x = 55$

$$^{53}_{24}\text{Cr} + {}^{4}_{2}\text{He} \rightarrow \underline{\quad} + 2\,{}^{1}_{0}\text{n}$$

> The proton number is determined from the sum of the proton numbers of the reactants and products.
> $24 + 2 = x + 0$ thus $x = 26$

> Fe has 26 protons.
> The isotope of Fe with a mass of 55 is ^{55}Fe.

INSIGHT:	In all nuclear reactions and radioactive decays the following rules are obeyed:
	1) The sum of the mass numbers of the reactants must equal the sum of the mass numbers of the products.
	2) The sum of the proton numbers of the reactants must equal the sum of the proton numbers of the products.

Half-life Sample Exercises

5. *Tritium, ^{3}H, a radioactive isotope of hydrogen has a half-life of 12.26 y. If 2.0 g of ^{3}H are made, how much is left 36.78 y later?*

The correct answer is: 0.25 g.

$$2.0 \text{ g of } {}^{3}\text{H} \xrightarrow{12.26\text{ y}} 1.0 \text{ g of } {}^{3}\text{H in the 1}^{\text{st}} t_{1/2}$$

$$1.0 \text{ g of } {}^{3}\text{H} \xrightarrow[24.52\text{ y total}]{\text{another 12.26 y}} 0.50 \text{ g of } {}^{3}\text{H in the 2}^{\text{nd}} t_{1/2}$$

$$0.50 \text{ g of } {}^{3}\text{H} \xrightarrow[36.78\text{ y total}]{\text{another 12.26 y}} 0.25 \text{ g of } {}^{3}\text{H in the 3}^{\text{rd}} t_{1/2}$$

6. A sample of a radiopharmaceutical containing 10.0 mg of ^{15}O is injected into a patient. A few minutes later the patient's body contains 0.625 mg of ^{15}O remaining from the sample. How long has the radiopharmaceutical been inside the patient? The half-life of ^{15}O is 2.03 minutes.
        The correct answer is: 8.12 minutes

0.625 mg of radiopharmaceutical remaining in the patient from the original 10.0 mg indicates that one-sixteenth of the sample remains, $\left(\dfrac{0.625 \text{ mg}}{10.0 \text{ mg}} = 0.0625 = \dfrac{1}{16} \right)$. To decrease the original sample by one-sixteenth the sample must have passed through four half-lives, $\left(\dfrac{1}{2} \times \dfrac{1}{2} \times \dfrac{1}{2} \times \dfrac{1}{2} = \dfrac{1}{16} \right)$. Thus the radiopharmaceutical has been inside the patient for 4×2.03 minutes = 8.12 minutes .

Fission Reaction Sample Exercise
7. What is the missing product nucleus of this fission reaction?
$$^{239}Pu + n \rightarrow\,^{100}Nb + ? + 2n$$
Correct answer is ^{138}I.

> The mass number of the missing product is determined from the sum of the mass numbers of the reactants and products.
> $239 + 1 = 100 + x + 2$ thus $x = 138$

$$^{239}_{94}Pu \;+\; ^{1}_{0}n \;\rightarrow\; ^{100}_{41}Nb + ? + 2\,^{1}_{0}n$$

> The proton number is determined from the sum of the proton numbers of the reactants and products.
> $94 + 0 = 41 + x + 2(0)$ thus $x = 53$

> I has 53 protons.
> The isotope of I with a mass of 138 is ^{138}I.

INSIGHT: In **fission** reactions, a *more massive nucleus is split into two lighter nuclei.* A representation of a chain fission reaction showing four completed fission reactions is shown below.

140

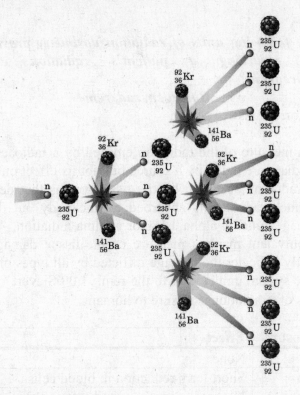

Fusion Reaction Sample Exercise

8. *What is the missing product nucleus of this fusion reaction?*

$$^4He + {}^3H \rightarrow ? + n$$

The correct answer is ^6Li.

Again, the mass number of the missing product is determined from the sum of the mass numbers of the reactants and products.
$4 + 3 = x + 1$ thus $x = 6$

$$^4_2He \; + \; {}^3_1H \; \rightarrow ? \; + \; {}^1_0n$$

The missing proton number is determined from the sum of the proton numbers of the reactants and products.
$2 + 1 = x + 0$ thus $x = 3$

Li has 3 protons.
The isotope of Li with a mass of 6 is ^6Li.

INSIGHT: In **fusion** reactions, *two lighter nuclei are merged into a more massive nucleus.* The reaction above indicates that a ^4He and ^3H merge into a ^6Li nucleus. This is one possible fusion reaction that occurs in stars.

141

Radiation Dosimetry

9. Which of the following units of radiation dosimetry provides the most useful information regarding a patient's radiation exposure from a radiopharmaceutical?

Roentgen, rad, rem

The correct answer is rem.

The Roentgen, a measure of the radiation emitted by a radioactive source, indicates the amount of radiation emitted by a source but not its effect on a human. The rad, a measure of the radiation absorbed by the tissue versus that delivered to the tissue, indicates how much radiation is absorbed by the body but does not differentiate between the damage done by alpha, beta, or gamma radiation. The rem (an acronym for <u>R</u>oentgen <u>e</u>quivalent <u>man</u>), a measure of the tissue damage done by radiation exposure, correctly indicates the damage inflicted by all types of radiation. A Sievert (Sv) is the metric system unit similar to the rem. 1.00 Sievert = 100 rem. This table shows the effects of γ radiation exposure to humans.

Radiation Exposure	Effect
< 25 rem	None
25 to 100 rem	Short-term reduction in blood cells
100 – 200 rem	Nausea, fatigue, vomiting for > 125 rem, longer term blood cell reduction
200 - 300 rem	First day nausea and vomiting, two weeks later appetite loss, sore throat, diarrhea, death for 10-35% of humans within 30 days, remaining recover in ~ 3 months
300 - 600 rem	Within hours nausea, vomiting, and diarrhea, one week later bleeding and inflammation of mouth and throat, for > 450 rems 50% of people die
> 600 rem	Within hours nausea, vomiting, and diarrhea, followed by death in two weeks for nearly 100% of humans

Math Review

Introduction

Introductory chemistry classes require some basic mathematical skills. These include many which you were taught earlier in your academic career but have probably forgotten from lack of use. In this section we will address those rusty math skills so that you can call upon them as necessary during your study of general chemistry. The important topics to learn from this section are: a) proper use of scientific notation, b) basic calculator skills including entering numbers in scientific notation, c) how to round off numbers, d) use of the quadratic equation, e) the Pythagorean theorem, and f) some rules of logarithms. Most of these skills will be required in some chapters of your introductory chemistry course.

Math Review Key Equations & Concepts

1. $$x = \frac{-b \pm \sqrt{b^2 - 4ac}}{2a}$$

   This equation is used to determine the solutions to quadratic equations, i.e. equations of the form $ax^2 + bx + c$. You will frequently encounter quadratic equations in equilibrium problems.

2. $x = a^y$ then $y = \log_a x$

   $$\log(x \cdot y) = \log x + \log y$$

   $$\log\left(\frac{x}{y}\right) = \log x - \log y$$

   $$\log(x^n) = n \log x$$

   The first equation is the definition of logarithms. The other equations are basic rules of algebra using logarithms. These rules apply to logarithms of any base, including base e or natural logarithms, ln. These equations will be used frequently in kinetics and thermodynamic expressions.

Scientific and Engineering Notation

In the physical and biological sciences it is frequently necessary to write numbers that are extremely large or small. It is not unusual for these numbers to have 20 or more digits beyond the decimal point. For the sake of simplicity and to save space when writing, a compact or shorthand method of writing these numbers must be employed. There are two possible, equivalent methods called either scientific or engineering notation. In both methods the insignificant digits that are placeholders between the decimal place and the significant figures are expressed as powers of ten. Significant digits are then multiplied by the appropriate powers of ten to give a number that is both mathematically correct and indicative of the correct number of significant figures to use in the problem. To be strictly correct, the significant figures should be between 1.000 and 9.999, however this particular rule is frequently ignored and, in fact, must be ignored when adding numbers in scientific notation that have different powers of ten. The only difference between scientific and engineering notation is how the powers of ten are written. Scientific notation uses the symbolism $x \: 10^y$ whereas engineering notation uses the symbolism Ey. Engineering notation is frequently used in calculators and computers.

INSIGHT:

> *Positive powers of ten* indicate that the decimal place has been *moved to the left that number of spaces*.
> *Negative powers of ten* indicate that the decimal place has been *moved to the right that number of spaces*.

A few examples of both scientific and engineering notation are given in this table.

Number	Scientific Notation	Engineering Notation
10,000	1×10^4	1E4
100	1×10^2	1E2
1	1×10^0	1E0
0.01	1×10^{-2}	1E-2
0.000001	1×10^{-6}	1E-6
23,560	2.356×10^4	2.356E4
0.0000965	9.65×10^{-5}	9.65E-5

It is important for your success in chemistry that you understand how to use both of these methods of expressing very large or small numbers. Familiarize yourself with both methods.

Basic Calculator Skills

Introductory chemistry courses require calculations that are frequently performed on calculators. You do not need to purchase an expensive calculator for your course. Rather you need a calculator that has some basic function keys. Common important functions to look for on a scientific calculator are log and ln, antilogs or 10^x and e^x, ability to enter numbers in scientific or engineering notation, x^2, $1/x$, $\sqrt{}$ or multiple roots, like a cube or higher root. More important than having an expensive calculator is knowing how to use your calculator. It is strongly recommended that you study the manual that comes with your calculator and learn the basic skills of entering numbers and understanding the answers that your calculator provides. For a typical Introductory chemistry course there are three important calculator skills that you must be proficient with.

1) Entering Numbers in Scientific Notation

Get your calculator and enter this number into it, 2.54×10^5. The correct sequence of strokes is press 2, press the . button, press 5, press 4, and *then press either EE, EX, EXP or the appropriate exponential button on your calculator*. **Do not press x 10 before you press the exponential button!** This is a very common mistake and will cause your answer to be 10 times too large.

> After you have entered 2.54×10^5 into your calculator, press the Enter or = button and look at the number display. If it displays 2.54E6 or 2.54×10^6, you have mistakenly entered the number. Correct your number entering method early in the course before it becomes a bad habit causing you to miss many problems.

2) Taking Roots of Numbers and Entering Powers

Frequently we must take a square or cube root of a number to determine the correct answer to a problem. Most calculators have a square root button, $\sqrt{}$. To take square roots just enter the number into your calculator and press the $\sqrt{}$ button to get your answer. For example, take the square root of 72, your answer should be 8.49. Some, but not all, calculators have a $\sqrt[3]{}$ button as well. If your calculator does not have a $\sqrt[3]{}$ button then you can use the y^x button to achieve the same result. To take a cube root, enter 1/3 or 0.333 as the power and the calculator will take a cube root for you. For example, enter $27^{0.333}$ into your calculator. You should get 3.00 as the correct answer. If you need a fourth root, enter ¼ which is 0.25 as the power, and so forth for higher roots.

3) Taking base 10 logs and natural or naperian logs, ln

Many of the functions in thermodynamics, equilibrium, and kinetics require the use of logarithms. All scientific calculators have log and ln buttons. To use them simply enter your number and press the button. For example, the log 1000 = 3.00 and the ln of 1000 = 6.91.

INSIGHT:	A common mistake that students frequently make is taking the ln when the log is needed and vice versa. Be careful which logarithm you are calculating for the problem.

Rounding of Numbers

When determining the correct number of significant figures for a problem it is frequently necessary to round off an answer when truncating to the appropriate number of significant figures. Basically, if the number immediately after the last significant figure is a 4 or lower, round down. If it is a 6 or higher, round up. The confusion arrives when the determining number is a 5. If the following number is a 5 followed by a number greater than zero, round the number up. If the number after the 5 is a zero, then the textbook used in your course will have a rule based upon whether the following number is odd or even. You should use that rule to be consistent with your instructor. The following examples illustrate these ideas. In each case the final answer contains three significant figures.

Initial Number	**Rounded Number**
3.67492	3.67
3.67623	3.68
3.67510	3.68
3.67502	Use your textbook rule.

Use of the Quadratic Equation

Equilibrium problems frequently require solutions of equations of the form $ax^2 + bx + c$. These are quadratic equations and the two solutions can always be determined using this formula.

$$x = \frac{-b \pm \sqrt{b^2 - 4ac}}{2a}$$

For example, if the quadratic equation to be solved is $3x^2 + 12x - 6$, then a = 3, b = 12, and c = -6. The two solutions can be found in this fashion.

$$x = \frac{-b \pm \sqrt{b^2 - 4ac}}{2a}$$

$$x = \frac{-12 \pm \sqrt{12^2 - 4(3)(-6)}}{2(3)}$$

$$x = \frac{-12 \pm \sqrt{144 + 72}}{2(3)}$$

$$x = \frac{-12 \pm \sqrt{216}}{6}$$

$$x = \frac{-12 \pm 14.7}{6} = \frac{2.7}{6} \text{ and } \frac{-26.7}{6}$$

$$x = 0.45 \text{ and } -4.45$$

> **INSIGHT:** Quadratic equations always have two solutions. In equilibrium problems one of the solutions will not make physical sense. For example, it will give a negative concentration for the solutions or produce a concentration that is outside the possible ranges of solution concentrations. It is your responsibility as a student to choose the correct solution based on your knowledge of the problem.

Rules of Logarithms

Logarithms are convenient methods of writing numbers that are exceptionally large or small and expressing functions that are exponential. They also have the convenience factor of making the multiplication and division of numbers written in scientific notation especially easy because in logarithmic form addition and subtraction of the numbers is all that is required. By definition, a logarithm is the number that the base must be raised to in order to produce the original number. For example, if the number we are working with is 1000 then 10, the base, must be cubed, raised to the 3rd power, to reproduce it. Mathematically, we are stating that $1000 = 10^3$, so the log (1000) = 3. There are three commonly used rules of logarithms that you must know. They are given below.

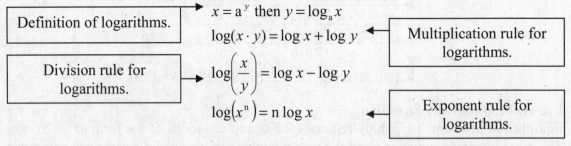

Definition of logarithms.	$x = a^y$ then $y = \log_a x$	
Division rule for logarithms.	$\log(x \cdot y) = \log x + \log y$	Multiplication rule for logarithms.
	$\log\left(\frac{x}{y}\right) = \log x - \log y$	
	$\log(x^n) = n \log x$	Exponent rule for logarithms.

> **INSIGHT:** These rules are correct for base 10, natural, or any other base logarithms.

Significant Figures for Logarithms

When we write this number, 2.345×10^{12}, we understand that there are 4 significant figures (the 2, 3, 4, and 5) but the power of 10 (the number 12) is not counted as significant. If we take the log of 2.345×10^{12} the number of significant figures must remain the same, 4. The log of $2.345 \times 10^{12} = 12.3701...$ What numbers indicate the exponents that are present in scientific notation? In logarithms, the numbers to the left of the decimal place, called the characteristic, are insignificant and the ones to the right, the mantissa, are significant. Thus the log $(2.345 \times 10^{12}) = 12.3701$ and both numbers have 4 significant figures.

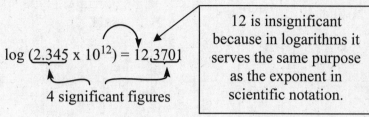

$$\log (\underline{2.345} \times 10^{12}) = 12.\underline{3701}$$

4 significant figures

12 is insignificant because in logarithms it serves the same purpose as the exponent in scientific notation.